Patrick Maaß

Multiplexing of PCR Assays in Breast Cancer Analysis

Bachelor + Master
Publishing

Maaß, Patrick: **Multiplexing of PCR Assays in Breast Cancer Analysis**, Hamburg,
Diplomica Verlag GmbH 2011
Originaltitel der Abschlussarbeit: Multiplexing of Singleplex Real-Time-PCR Assays into
Duplexes and Triplexes

ISBN: 978-3-86341-107-7
Druck: Bachelor + Master Publishing, ein Imprint der Diplomica® Verlag GmbH,
Hamburg, 2011
Zugl. Johannes Gutenberg-Universität Mainz, Mainz, Deutschland, Bachelorarbeit, 2009

Bibliografische Information der Deutschen Nationalbibliothek:
Die Deutsche Nationalbibliothek verzeichnet diese Publikation in der Deutschen
Nationalbibliografie;
detaillierte bibliografische Daten sind im Internet über http://dnb.d-nb.de abrufbar.

Die digitale Ausgabe (eBook-Ausgabe) dieses Titels trägt die ISBN 978-3-86341-607-2
und kann über den Handel oder den Verlag bezogen werden.

Dieses Werk ist urheberrechtlich geschützt. Die dadurch begründeten Rechte,
insbesondere die der Übersetzung, des Nachdrucks, des Vortrags, der Entnahme von
Abbildungen und Tabellen, der Funksendung, der Mikroverfilmung oder der
Vervielfältigung auf anderen Wegen und der Speicherung in Datenverarbeitungsanlagen,
bleiben, auch bei nur auszugsweiser Verwertung, vorbehalten. Eine Vervielfältigung
dieses Werkes oder von Teilen dieses Werkes ist auch im Einzelfall nur in den Grenzen
der gesetzlichen Bestimmungen des Urheberrechtsgesetzes der Bundesrepublik
Deutschland in der jeweils geltenden Fassung zulässig. Sie ist grundsätzlich
vergütungspflichtig. Zuwiderhandlungen unterliegen den Strafbestimmungen des
Urheberrechtes.

Die Wiedergabe von Gebrauchsnamen, Handelsnamen, Warenbezeichnungen usw. in
diesem Werk berechtigt auch ohne besondere Kennzeichnung nicht zu der Annahme,
dass solche Namen im Sinne der Warenzeichen- und Markenschutz-Gesetzgebung als frei
zu betrachten wären und daher von jedermann benutzt werden dürften.

Die Informationen in diesem Werk wurden mit Sorgfalt erarbeitet. Dennoch können
Fehler nicht vollständig ausgeschlossen werden, und die Diplomarbeiten Agentur, die
Autoren oder Übersetzer übernehmen keine juristische Verantwortung oder irgendeine
Haftung für evtl. verbliebene fehlerhafte Angaben und deren Folgen.

© Bachelor + Master Publishing, ein Imprint der Diplomica® Verlag GmbH
http://www.diplom.de, Hamburg 2011
Printed in Germany

Table of Contents

1		Introduction	1
	1.1	Breast Cancer	1
	1.2	RNA Expression Profiling and Prediction	2
	1.3	Conceptional Formulation and Objectives	3
2		Materials and Methods	5
	2.1	Materials	5
	2.1.1	Instruments and Consumables	5
	2.1.2	Buffers and Chemicals for Nucleic Acid Purification	5
	2.1.3	Nucleic Acids, Cell Lines and Tissue Samples	5
	2.1.4	Synthetic Oligonucleotides	6
	2.1.5	Chemicals and Kits for PCR	6
	2.1.6	Software	6
	2.2	Methods	7
	2.2.1	General Approach	7
	2.2.2	Purification of Nucleic Acids	7
	2.2.3	Creation of a Standard Reference RNA Pool	8
	2.2.4	Real-Time Kinetic RT-PCR	9
	2.2.5	Designing Dually Labeled Primer-Probe Sets	10
	2.2.6	Dilution of Primer-Probe Sets	11
	2.2.7	Setup of Singleplex kPCR Assays	11
	2.2.8	Evaluation of Different Reporter Dyes	12
	2.2.9	Controlling of Primer-Probe-Set Performance	13
	2.2.10	Testing Combinations of Two Sets (Duplexing)	14
	2.2.11	Testing Combinations of Three Sets (Triplexing)	16
3		Results	18
	3.1	Identification of Suitable Reporter Dyes	18
	3.2	Evaluation of Primer-Probe-Performance	23
	3.3	Identification of Suitable Duplex Combinations	24
	3.4	Identification of Suitable Triplex Combinations	27
4		Discussion	32
	4.1	Statistical Evaluation of Primer Performances	32
	4.1.1	Slope	32
	4.1.2	Efficiency	33
	4.1.3	Y-Intercept	35
	4.2	Suitable Reporter Dyes and Quencher	36
	4.3	Optimization of Assays and kinetic RT-PCR Parameter	37
5		Summary	40
6		Literature	
7		Appendix	
	7.1	Abbreviations	
	7.2	Figures	
	7.3	Tables	
	7.4	Oligonucleotide Sequences	
	7.5	Origins of Breast Cancer Samples for MAVPOOL080623a	
	7.6	Acknowledgements	

1 Introduction

1.1 Breast Cancer

"Cancer" describes a group of various diseases, where cells start changing their molecular structure and begin to grow and to supersede normal cells. Cancer is induced by numerous different elicitors, which finally all lead to an interference of the genetically regulated balance between cell cycle and apoptosis. Although every organ in the human body can be afflicted with cancer, there are significant differences in frequency relating amongst others to age, sex, geographic region and personal habits.

In the industrialized countries, breast cancer is the leading cause of death for women at the age between 30 and 60 years. With estimated 636.000 incident cases in the developed countries and 514.000 in the developing countries, breast cancer is the most prevalent cancer type among woman worldwide [World Cancer Report 2008].

Once detected, the cancer is classified based upon pathological characterizations of the tumor or a biopsy and the lymph nodes. A clinical way of characterizing the tumor is the TNM-classification, which describes the size of the tumor (T), the number of affected lymph nodes (N) and the existence of distant metastases (M). The histological classification characterizes the carcinoma according to its structural and cellular appearance and the amitosis rate leading to a grading from 1 to 3. An immuno-histological examination provides information about the estrogen- and the progesterone-receptor- and about the Her-2/neu-status [Wolff A et al 2007].

Breast cancer is a very heterogeneous disease. There are basic classifications that are unquestioned, even today. Recent studies confirmed the need to determine well known markers (i.e. estrogen (ER) and progesterone (PR) receptor status [Garcia-Closas M, 2008] or HER2 status [Science Daily, 2007]), but the large variety of subtypes and the corresponding different molecular pattern impede a uniform treatment.

Although already today other factors than anatomical classifications are being taken into consideration, there exists a need for further biological markers to assist the physician in charge with his evaluation. Beside the diagnostic recognition, the choice of appropriate therapy and the prediction of prognosis are goals that should be reached in order to prevent early stage cancer patients from therapies that provide minimal benefit but reduce their quality of life by intense adverse reactions [Ganz PA et al 2004].

1.2 RNA Expression Profiling and Prediction

Already today there are numerous genes associated with breast cancer occurrence, therapy selection and prognosis [van 't Veer LJ et al 2002]. These profiles can be generated by different techniques, like DNA-microarrays or kinetic polymerase chain reaction (kPCR). The kPCR displays the handier platform for research laboratories in terms of choices of genes. It was the method of choice during this work.

Besides choosing the appropriate genes for analysis, there are several other elementary requirements that may lead to solid results like a statistical relevant number of tissue samples and a dependable and accurate follow-up documentation of the patient's disease history. Within this follow-up, clinical parameters and important events like the recurrence of tumors, discovery of metastasis or demise must be documented. These parameters can then be correlated to gene expression profiles by extensive statistical analysis [Buyse M et al 2007].

Targeting patients with lymph-node-negative breast cancer, Siemens Healthcare Diagnostics designed a diagnostic assay to predict the probability of developing distant metastasis after surgery within 5 to 10 years. This kPCR-based quantification method draws upon several previously determined and selected genes, and uses formalin-fixed, paraffin-embedded (FFPE) tissue samples of 310 patients from the Department of Obstetrics and Gynecology, Johannes-Gutenberg-University Mainz, Germany [Stropp U et al 2008]. After quantifying these genes of interest (GOI) and normalizing on housekeeping genes (HKG), the determined amounts were correlated with comprehensive statistical data of the patients. Finally an algorithm was developed to predict the forecast of patients and to determine their need for chemotherapy.

This diagnostic assay consists of 9 GOIs, 2 HKGs and 1 DNA control and is performed within a 96-well polypropylene plate.

To verify suitable algorithms, they have to be confirmed within control groups. 7 different algorithms were developed by in-house statisticians, who applied different mathematical approaches. These 7 algorithms were then applied to control groups to determine the most reliable and reproducible method.

1.3 Objectives

The goal of this work was to reduce the amounts of wells per patient and assay by transforming the existing 12 singleplex assays into duplex- and triplex-formats, in order to increase the number of samples per plate or to allow more reference-control genes within a multiplex assay.

The primary goals of this project are:

- Reduction of cost of approximately 50% for mastermixes per patient
- Higher throughput due to larger sample number per plate
- Larger number and higher variety of GOIs per patient and plate
- Retrenchment of very valuable RNA material
- Generation of resources to run additional quality assurance (housekeeping-genes)

The platform the kPCRs were performed on was Stratagene's MX3005p – a multichannel kPCR machine which possesses a tungsten halogen lamp and 5 different filter sets for parallel analysis of the corresponding amount of reporter dyes. It is able to excite and detect fluorophores with an excitation and emission wavelength between 400 and 700nm [Marras 2005]. The identification of suitable reporter dyes for the filter sets by performance-analysis was the initial requirement during this work, following the evaluation of compatibility of dyes among each other within multiplex assays. Finally the most promising and reliable multiplex assays had to be identified amongst the numerous different possible combinations.

In the context of this evaluation process great attention was put on consistent parameter during kPCR (i.e. buffer conditions, reaction volume, usage of identical lots, cycle conditions, etc.) and to strictly exclude any crosstalk between the individual channel.

The results emerging from this work will deliver a substantiated basis upon which further comparative studies with more comprehensive numbers of patients can be measured.

Then, in-house bioinformaticians will test the existing algorithm on data gained from multiplex assays of a statistical larger universe of patients.

2 Materials and Methods

2.1 *Materials*

2.1.1 Instruments and Consumables

<u>EPPENDORF, Hamburg, Germany</u>
Centrifuge 5804, Cat. No.: 5804 000.013 with
Centrifuge Rotor A-2-DWP, Cat. 5804 740.009
Pipet Reference variable 0,1 – 2,5µl, Cat. No.: 4910 000.085
Pipet Reference variable 0,5 – 10µl, Cat. No.: 4910 000.018
Pipet Reference variable 10 – 100µl, Cat. No.: 4910 000.042
Pipet Reference variable 100 – 1000µl, Cat. No.: 4910 000.069

<u>GILSON. Middleton, WI, USA</u>
Repititive Pipet Distriman. Cat. No.: F164001

<u>HEIDOLPH, Schwabach, Germany</u>
Vortexer Reax Control, Cat. No.: 541-11000

<u>LABNET, Woodbridge, NJ, USA</u>
Centrifuge Quick Spin Minifuge, Cat. No.: C1201

<u>SARSTEDT, Siegburg, Germany</u>
Micro tube with screw cap 1.5ml, Cat. No.: 72.692.005

<u>STRATAGENE, La Jolla, CA, USA</u>
Mx3005P QPCR System, Cat. No.: 401458 with Alexa405-, ROX-, HEX-, FAM- and Cy5-Filter
Mx3000P®/Mx3005P® Optical Strip Caps, Cat. No.: 401425
Mx3000P® 96-well-plates skirted, Cat. No.: 401334

<u>TECAN, Crailsheim, Germany</u>
Robot Genesis Workstation 150

2.1.2 Buffers and Chemicals for Nucleic Acid Purification

<u>QIAGEN, Hilden, Germany</u>
Proteinase K, Cat. No.: 19133

<u>SIEMENS MEDICAL SOLUTIONS DIAGNOSTICS GMBH, Eschborn, Germany</u>
Lysis Buffer, Cat. No.: 03745099
Washing Buffer I, Cat. No.: 03745226
Washing Buffer II, Cat. No.: 03746737
Washing Buffer III, Cat. No.: 03742146
Elution Buffer, Cat. No.: 03742677
Magnetic Beads, Cat. No.: 03749787

2.1.3 Nucleic Acids, Cell Lines & Tissue Samples

<u>DSMZ, Braunschweig, Germany</u>

German Collection of Microorganisms and Cell Cultures, Braunschweig
MCF-7 cell line, Cat.No.: DSMZ ACC 115
(RNA isolated using QIAGEN, RNeasy Mini Kit, Cat.No. 74104 according to manufacturers instructions)

STRATAGENE, La Jolla, CA, USA
Stratagene QPCR Reference Total RNA, Human, Cat. No.: 750500

BREAST CANCER SAMPLES
Prof. Dr. med Stephan Störkel, Helios Klinikum Wuppertal, Departement for Pathology
Prof. Dr. Med Carsten Denkert, Charite Berlin, Departement for Pathology
Samples from women with invasive breast cancer, surgery 2003
5-10µm Slides of FFPE Samples with tumor cell content >30%
Informed Consent of Patients at hand.

2.1.4 Synthetic Oligonucleotides

MICROSYNTH, Balgach, Switzerland
HPLC-purified oligonucleotides and MALDI/TOF-controlled probes
(See appendix for complete listing of oligonucleotide sequences)

2.1.5 Chemicals and Kits for PCR

AMBION, Austin, TX, USA
Nuklease-free water (not DEPC-treated), Cat. No.: AM9932
RNaseZap, Cat. No.: AM9782
DNAzap, Cat. No.: AM9890

INVITROGEN, Carlsbad, CA, USA
SuperScript III Platinum One-Step qRT-PCR kit, Cat. No.: 11732-088

2.1.6 Software

MICROSOFT, Redmont, WA, USA
Microsoft Office Professional Edition 2003

APPLIED BIOSYSTEMS, Foster City, CA, USA
Primer Express Version 2.0.0

STRATAGENE, La Jolla, CA, USA
MxPro – Mx3005p v4.01 Build 369, Scheme 80

2.2 Methods

2.2.1 General Approach

All working steps concerning the handling of RNA were performed at a reserved working place and with instruments, which were kept RNAse free.

Unless otherwise noted, when *water* is mentioned in the text, nuclease-free water is referred to.

2.2.2 Purification of Nucleic Acids

Nucleic acids (NA) used in the context of this work were derived from formalin-fixed, paraffin embedded (FFPE) tissues of women with breast cancer surgery and from MCF-7 cell lines (see appendix 6.2). The tissue samples were available as paraffin sections of 5-10μm thickness and were stored at 8°C in 1,5ml Sarstedt reaction tubes. MCF-7 cell line is commercially available from the German Collection of Microorganisms and Cell Cultures, Braunschweig.

All nucleic acids extracted from patients breast-cancer samples were purified by an automated in-house sample preparation method using magnetic beads [Hennig G and Hildenbrand K 2006], specific solutions and a pipetting-robot (Figure 1).

Within this method, RNA was extracted from FFPE-samples. For this, samples were treated with a lysis buffer and Proteinase K and incubated for 2h at 65°C.

Together with a special binding buffer, the magnetic beads were added and incubated for another 10min at room temperature (RT).

While magnetizing the bead-bounded NAs, three wash-steps with washing buffer I - III removed waste compounds.

Finally an elution buffer was used to separate the NAs from the magnetic beads at 70°C and a DNAse I digestions removed the DNA to obtain the desired, pure RNA.

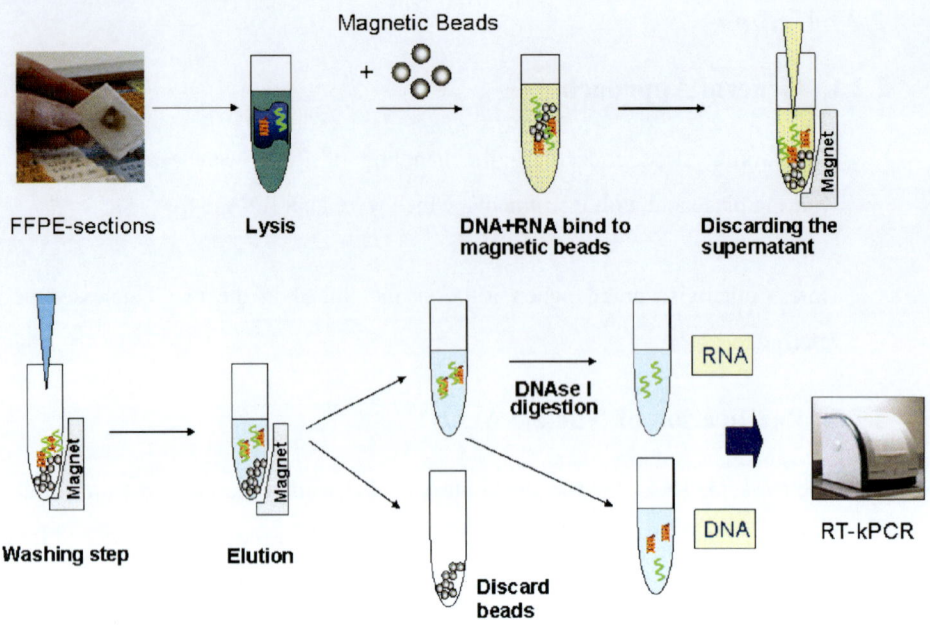

Figure 1: Schematic Workflow of Nucleic Acid Extraction from FFPE Samples

2.2.3 Creation of a Standard Reference RNA Pool

A standard reference RNA pool was needed for further testings. Previous experiments had shown, that commercially available reference RNAs (i.e. Universal Human Reference RNA, Stratagene, Cat. No.: 740000) did not represent a realistic gene profile, since some genes of interest (GOI) are only amplified in patients RNA but not in commercial reference RNAs. (Mojica WD, Stein L, Hawthorn L, 2008).

Therefore, RNA from 83 breast cancer samples and MCF-7 cells (see appendix for details of origin) was isolated and the extracted RNA was combined into a common pool. A dilution series was made by diluting the RNA with nuclease free water according to the following scheme:

Table 1: Dilution Series of Human Breast Cancer Samples for Setting up a Standard Reference RNA Pool

RNA [µl]	H$_2$O [µl]		Dilution
5000,0	0,0		1:1
1250,0	3750,0	⇒	1:4
312,5	4687,5		1:16
78,1	4921,9		1:64

5000µl of each dilution was then aliquoted in 50 x 100µl and stored as "*MAVPOOL080623a*" at -80°C.

2.2.4 Real-Time Kinetic RT-PCR

The real-time quantitative RT-PCR method used to quantify RNA in breast cancer tissue combines two successive steps.

First, RNA is transcribed into cDNA by the enzyme *reverse transcriptase*, an RNA-dependent DNA-polymerase, which was first discovered in retroviruses (Gilboa E et al 1979) and which is able to synthesize a RNA-DNA-hybrid-strand from a single-stranded RNA, degrade the residual RNA and complete the molecule into a double-stranded cDNA.

In the second step, the cDNA serves as a template for the following quantitative polymerase chain reaction (PCR). The PCR uses two sequence-specific oligonucleotides and a DNA-dependent polymerase to amplify a definite DNA segment (Mullis KB, Faloona FA 1987).

An improvement of the PCR is the real-time quantitative PCR, where a third oligonucleotide, a hybridization probe labeled with two different fluorescent dyes and located between the forward- and the reverse-primer, is used. Since one dye works as the reporter dye (i.e. FAM, Cy5, Yakima Yellow, etc.) and the other one as the corresponding quencher (like TAMRA, BHQ1, BHQ2, etc.), the quencher absorbs the emission of the reporter dye by fluorescent resonance energy transfer (FRET).

When this dual-labeled probe hybridizes with the template DNA, the 5'-3' nucleolytic activity of the polymerase degrades this probe resulting in a loss of quenching activity. Thus, a continuous increase of occurs during PCR. The amount of fluorescence at a given time point during PCR corresponds to the amount of PCR product. Since fluorescence is measured following each PCR cycle, it is possible to observe the amplification process "real-time" and to count back to the initial amount of cDNA (Heid CA, Stevens J, Livak K J et al.1996).

2.2.5 Designing Dually Labeled Primer-Probe Sets

Primer design was accomplished with the help of the software tool *Primer Express v2.0.0* from Applied Biosystems. Although this software was built for designing TaqMan® primer and probe sets, it delivers excellent results for other real-time applications such as the MX3005p. When choosing *TaqMan® Primer and Probe Design*, the software operates with predefined parameters using empirical rules to calculate optimal sequences based upon the input sequence. The most important parameters for the probe were:

- Amplicon size should range from 50 – 150 base pairs (bp)
- G/C content should be kept between 30% and 80%
- Avoiding repeats of identical bases – especially of Guanine
- The melting temperature should be between 68°C and 70°C
- No 5'-terminal Guanine
- Primers should be designed a close to the probe as possible

(Source: Primer Express Software Version 3.0 Getting Started Guide)

The corresponding forward- and reverse-primer were also automatically designed by that software and their melting temperature should have been about 10°C below the probe-temperature.

Although all samples were treated with DNAse, this digestion is very often imperfect [Wink, 2004]. The use of RNA-specific primer probe sets copes with that specific problem. In order to avoid amplification of genomic DNA, RNA-specific, intron-spanning [Freeman, 1999] primer-probe sets were designed if possible, based upon the cDNA sequence.

All primer-probe sets were ordered from Microsynth, Switzerland in 0.2µmol scale and HPLC purified.

2.2.6 Dilution of Primer-Probe Sets

Each set consisted of two standard oligonucleotides and one dual-labeled probe. Both, the unmodified and the modified oligonucleotides were first diluted to a final concentration of 100μM according to the documents provided by Microsynth. The working solution consisted of 50μl (each) forward and reverse primer and 25μl probe filled up with nuclease free water to a total volume of 1000μl. Thus, the working solution consisted of two unmodified oligonucleotides (5μM each) and one dual-labeled probe (2,5μM).

2.2.7 Setup of Singleplex kPCR Assays

The standard setup for a singleplex QPCR was based upon a total volume of 20μl. The following quantities of the constituents were used:

Table 2: Setup of a Standard kPCR Assay

5,6	μl water
10,0	μl RT QPCR mastermix
1,0	μL MgSO4 (50mM)
1,0	μl primer-probe mix
0,4	μl RT/Taq mix
2,0	μl RNA
20μl	total

After combining mastermix, water, primer-probe mix and $MgSO_4$, the RT/Taq mix and the RNA was added, whereas both, the RT/Taq mix and the RNA must be handled on wet ice and all pipetting steps were performed on wet ice, too.

$MgSO_4$ might affect the performance of the PCR, since its concentration affects primer annealing, denaturation of the strands and product specifity. In addition, $MgSO_4$ is needed for the activity of enzymes. Since primer and nucleotides capture available $MgSO_4$ [Mülhardt, C., 2000] the concentration was elevated by adding 1μl of 50mM $MgSO_4$ to the mix. Since the 2x mastermix already contains 6mM $MgSO_4$ the final concentration is increased by 2.5mM to 5.5mM $MgSO_4$ [Henegariu O et al 1997].
After diluting the primer-probe mix, the final concentrations per assay are 250nM for each of the two unmodified oligonucleotides and 125nM for the dual-labeled probe.
Reactions were performed in 96-well polypropylene plates. After pipetting all required components, the plate is sealed with lids and centrifuged at 1500rpm for 5min before

transferring it to the Stratagene MX3005p QPCR machine. This step is useful to remove possibly existing bubbles on the ground of each well, resulting in better duplicates.

The thermal profile was set based upon Invitrogens recommendations regarding their SuperScript III Platinum One-Step Quantitative RT-PCR System protocol. An adjustment of 50 cycles instead of 40 was the only change that was carried out.

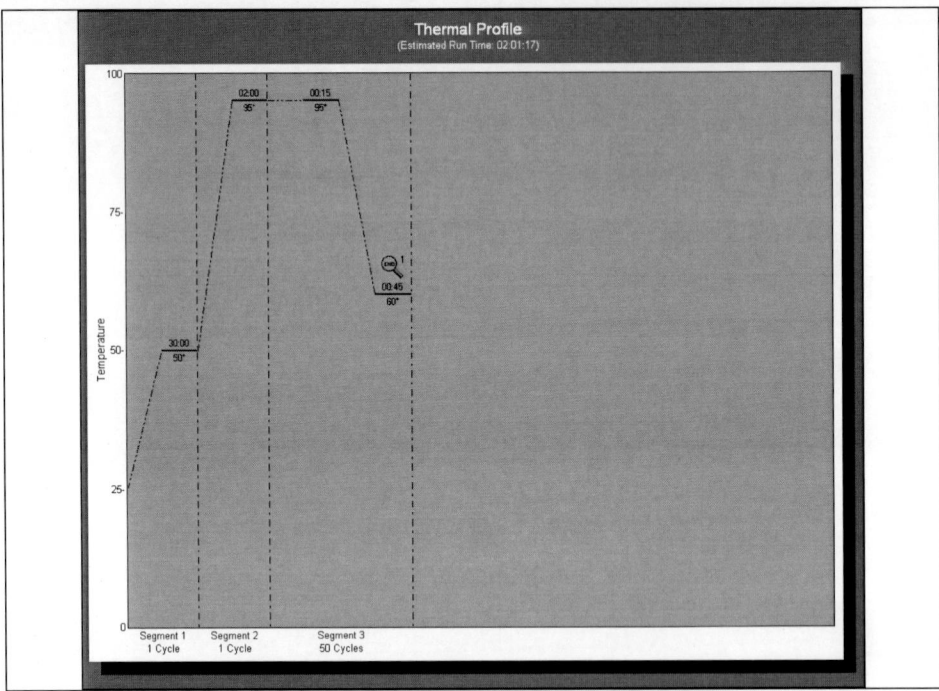

Figure 2: Thermal Profile of RT-QPCR on the Stratagene Mx3005p

Figure 2 shows the thermal profile during a RT-QPCR run, starting with a 30min incubation time at 50°C at which the cDNA synthesis is performed by the reverse transcriptase. Following this step, the real-time PCR is run. An initial 95°C step for 2:00min is followed by 50 cycles with 20sec at 95°C [denaturation of template DNA] and 45sec at 60°C [hybridization of template DNA and synthetic oligonucleotides]. Fluorescence measurement and data collection was performed during the hybridization of PCR products at the end of each cycle.

2.2.8 Evaluation of Different Reporter Dyes

The real-time QPCR-System Mx3005p from Stratagene is equipped with 5 different filters, which allow detection of up to 5 different dyes in one single reaction. The instrument used in the context of this work features the following filters:

- Alexa Fluor 350   (350nm ex. – 440nm em.)
- FAM / SYBR Green I   (492nm ex. – 516nm em.)
- HEX/JOE/VIC   (535nm ex. – 555nm em.)
- ROX / Texas Red   (585nm ex. – 610nm em.)
- Cy5   (635nm ex. – 665nm em.)

The Mx3005P system utilizes a quartz tungsten halogen bulb as its excitation light source. This bulb emits light from 350 to 750 nm (Stratagene, *Strategies*, 2005) and with that it is possible to test a large number of different dyes. The following different reporter dyes- and quencher-combinations were tested:

- FAM – BHQ1   (max Abs. 494nm – max. Em. 520nm)
- Yakima Yellow – BHQ1   (max Abs. 530nm – max. Em. 549nm)
- Cy5 – BHQ2   (max Abs. 646nm – max. Em. 662nm)
- Marina Blue – Dabcyl   (max Abs. 362nm – max. Em. 459nm)
- Alexa 405 – Dabcyl   (max Abs. 401nm – max. Em. 421nm)
- Pacific Blue – Dabcyl   (max Abs. 416nm – max. Em. 451nm)
- ATTO620 – BHQ2   (max Abs. 619nm – max. Em. 643nm)
- HEX – BHQ1   (max Abs. 535nm – max. Em. 556nm)
- Alexa 350 – Dabcyl   (max Abs. 346nm – max. Em. 442nm)
- ROX – BHQ2   (max Abs. 585nm – max. Em. 605 nm)
- ATTO550 – BHQ2   (max Abs. 554nm – max. Em. 576nm)

(Absorption and emission spectra derived from: *Custom Oligonucleotides*, Eurogentec, Belgium, 2007)

Before different dyes were tested for multiplexing, the compatibility with the Mx3005p QPCR machine and its filters had to be confirmed. For that purpose, all dyes had to pass an initial test, where they were tested on Stratagenes QPCR human Reference Total RNA and the in-house RNA pool "*MAVPOOL080623a* ". The test consisted of an amplification of a basic dilution series in which a broad spectrum of dilutions (1:1, 1:4, 1:16 and 1:64) was covered. Dyes, which were not detected by the system or produced very inconsistent results did not pass this initial test and were discarded.

2.2.9 Controlling of Primer-Probe-Set Performance

Before using new primer-probe sets in routine work, all sets undergo a process of testing their performance. Therefor each set is tested in the MAVPOOL080623a-dilution series and the performance is calculated (and, where applicable, compared to older charges).

Given dilutions (undiluted, 1:4, 1:16 and 1:64) and corresponding Ct-values make it possible, to calculate several parameters in Microsoft Excel:

- Slope calculated with the following formula:
  =**SLOPE(***Ct-Values***;LOG(***dilutions***))**
- Efficiency calculated with the following formula:
  =**10^(-1/***slope***)**
- Efficiency (%) calculated with the following formula:
  =**(10^(-1/***slope***)-1)*100**
- Coeff Determination calculated with the following formula:
  =**RSQ(***Ct-Values***;LOG(***dilutions***))**
- Y-Intersection calculated with the following formula:
  =**INTERCEPT(***Ct-Values***;LOG(***dilutions***))**

For evaluation purposes, the efficiency (%) is the most common value used to determine the quality of new, or the comparability of re-ordered primer-probe sets. The performance of new primer-probe sets should be between 95% and 105% in best cases; still well acceptable efficiency ranges are between 90% and 110%. For re-ordered primer-probe sets, an approximate difference lower than 5% should be obtained. In any case, only measured values within the linear range may be used for calculations. Experiments have shown that the linear range covers Ct-values between 18 and approx. 35. Dyes, that did not pass this performance test, were discarded.

2.2.10 Testing Combinations of Two Primer-Probe Sets (Duplexing)

Primer-probe sets that passed both, the performance- and the initial dye test, were tested against each other to examine their performance in duplex reactions. The remaining dyes FAM, Cy5, Yakima Yellow and ROX showed persuasive results and possess distinguishable absorption- and emission spectra (Figure 3 & 4).

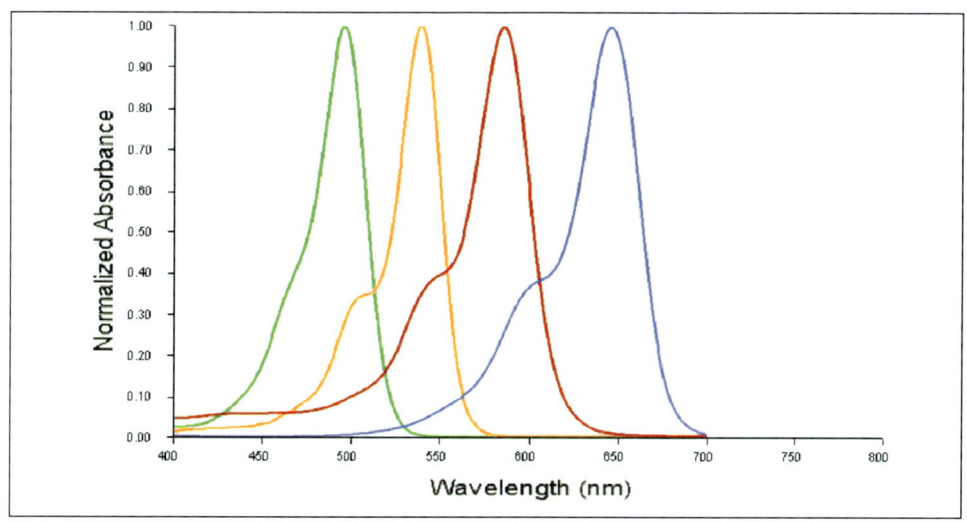

Figure 3: Absorption Spectra of FAM (green), Yakima Yellow (orange), ROX (red) and Cy5 (blue), Source: http://www.biosearchtech.com/hot/multiplexing.asp

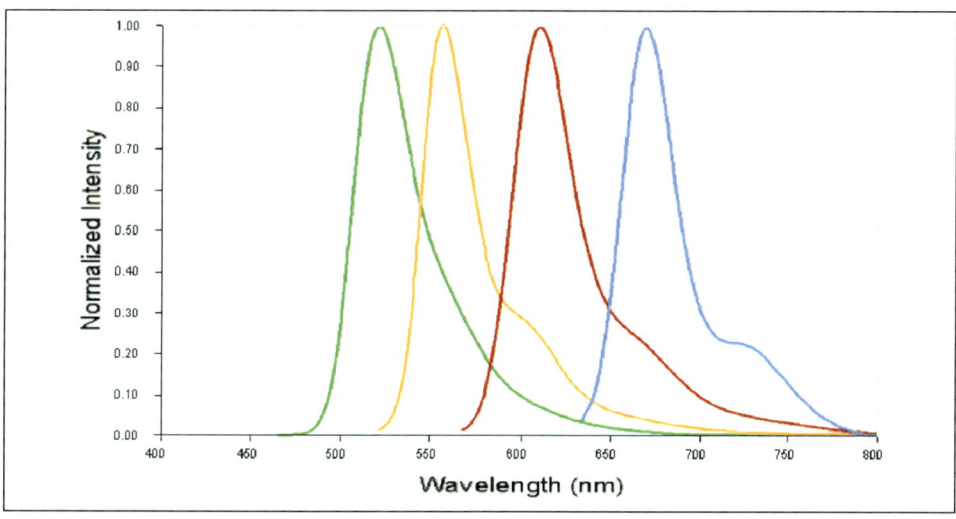

Figure 4: Emission Spectra of FAM (green), Yakima Yellow (orange), ROX (red) and Cy5 (blue), Source: http://www.biosearchtech.com/hot/multiplexing.asp

Thus, primer-probe sets were tested systematically against each other, whereas the individual oligonucleotide concentration and the total reaction volume were retained, but the pursuant amount of water was decreased. However, every duplex testing plate also included wells containing the corresponding singleplex assays of the two individual primer-probe sets. By choosing a specific fluorescence channel (i.e. FAM in an assay containing one FAM-labeled and one Cy5-labeled probe) it was possible to make a comparable performance-analysis of this specific probe within the mixture of two probes and to compare the ascending slope with the slope of the parallel singleplex assay.

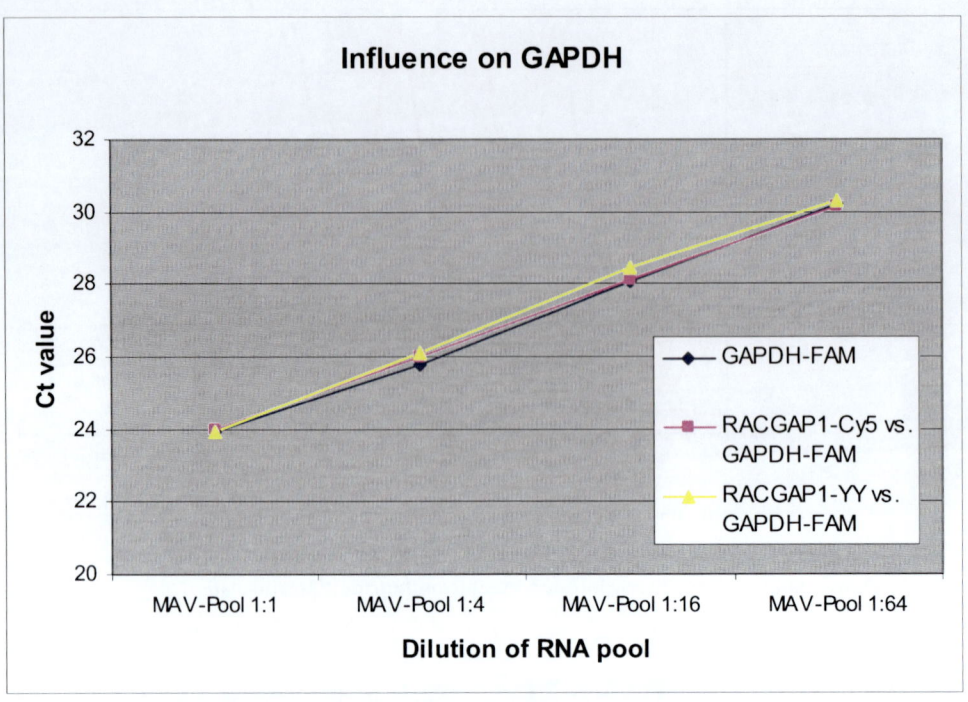

Figure 5: Comparative Evaluation of GAPDH-FAM in Duplex Reactions with RACGAP1-Cy5 (marked magenta) and with RACGAP1-YY (marked yellow) versus the Corresponding Singleplex Reaction of GAPDH-FAM (marked blue).

A discrepancy of the PCR efficiency of approx. 10% or more lead to a negative weighting of that individual assay (marked red in Table 3). Irregularly shaped curves led to markings (marked yellow in Table 3), whereas in most cases the duplex reactions worked fine (marked green in Table 3).

2.2.11 Testing Combinations of Three Sets (Triplexing)

Primer-probe sets that passed the duplex-evaluation were tested for their triplex capability. Similar to the duplex-testing, three different reporter dyes were combined into a single assay. Again, the individual oligonucleotide concentration and the total reaction volume were retained, but the pursuant amount of water was decreased. Analysis of the performance was made by plotting the Ct-values against the RNA-concentration and comparing the slope of the singleplex-assay against the slope of the triplex-assay while examining one discrete channel.

To prevent influences on the Ct-values caused by crosstalk between different filter-channels, great importance was attached to the fact, that no fluorescence signals could be detected in wells, where a specific dye was not present. For example: A well containing

FAM only must not produce values when analyzing this well with the Cy5- or the HEX-filter.

A third, even though somewhat imprecise attention was turned to the level of saturation. During duplex- and even more during triplex-assays, the individual reactions compete for the inserted components. It is therefore expected, that performance and saturation will deteriorate, but experiments have shown, that there are combinations, which show less influence than others. These combinations were favored for further testing.

3 Results

3.1 Identification of Suitable Reporter Dyes

The objective of this work was to combine singleplex kPCR assays of the following genes into multiplex assays:

Table 3: Overview of Tested Genes and Their Derivation

Gene Symbol	Definition
ESR1	Estrogen Receptor alpha
PGR	Progesteron Receptor
MLPH	Melanophilin
TOP2A	Topoisomerase II alpha
RACGAP1	Rac GTPase activating Protein 1
CHPT1	Choline Phosphotransferase 1
MMP1	Matrix Metallopeptidase 1
IGKC	Immunoglobulin kappa constant
CXCL13	Chemokine (C-X-C motif) Ligand 13
CALM2	Calmodulin 2
PPIA	Peptidylprolyl Isomerase A
PAEP	Progestagen-associated Endometrial Protein
GAPDH	Glyceraldehyde-3-Phosphate Dehydrogenase
OAZ1	Ornithine Decarboxylase Antizyme 1
ERBB2	v-erb-b2 erythroblastic leukemia viral oncogene homolog 2

All gene expression analysis performed within this work took place on Stratagenes MX3005p QPCR system. This system is equipped with 5 different filter sets, which make it possible to run multiplex analysis with up to 5 different reporter dyes.

To compare standard dyes that were recommended by Stratagene with none-proprietary alternative dyes, a set of four genes (TOP2A, RACGAP1, CHPT1 and IGKC) was chosen. Eleven different labeled sets per gene were then ordered to find the best dye that fits an individual filter setting.

To determine, whether a dye fits to a certain filter setting, a dilution series of Stratagenes QPCR human Reference Total RNA and the in-house RNA pool "*MAVPOOL080623a*" was quantified in a 3-fold assay. The analysis of those assays focused on the issue, whether a dye is detectable or not. For this purpose, the appearance of amplification curves, their shape and the plausibility of an appearance was considered.

Experiments showed, that Alexa 405, Alexa 350 and ATTO620 did not show any detectable fluorescence at all and were therefore discarded. Analysis of raw data showed very high background fluorescence, which was initially interpreted as a non-effective dye-linking to the oligonucleotide. This suspicion was later confirmed by the oligo supplier who analyzed his samples.

Pacific Blue and Marina Blue dyes had to be discarded, although they showed detectable fluorescence, but their emittance spectrum was not discrete enough. Marina Blue, a reporter dye with an absorption maximum of 362nm and an emission maximum of 459nm should have been detectable with the Alexa 350 filter. In fact, this dye was detectable (figure 6) but also showed signals in the FAM-filter (figure 7) and was therefore not practicable for multiplexing.

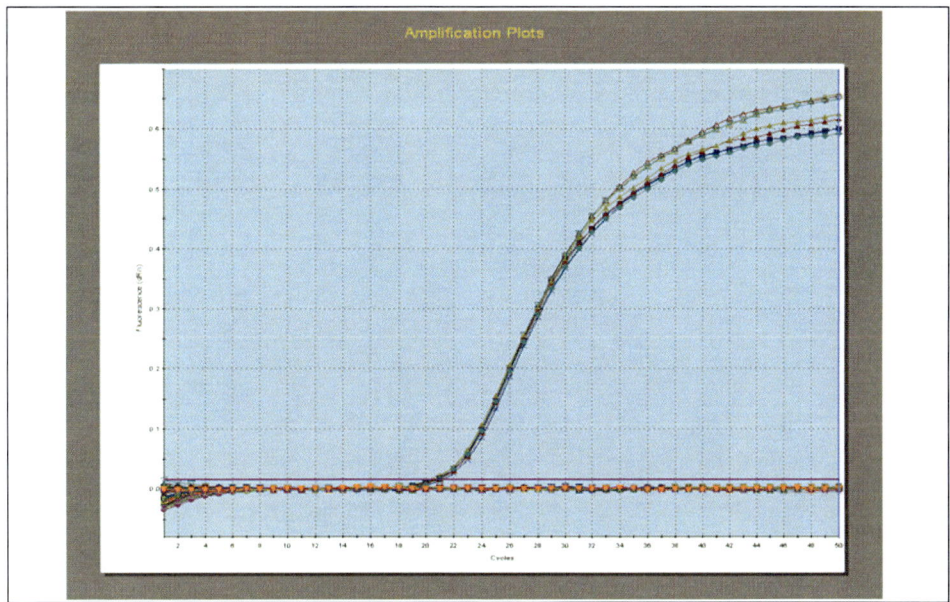

Figure 6: Fluorescence Signal of MLPH (marina blue labeled) in the Alexa 350 Filter Channel

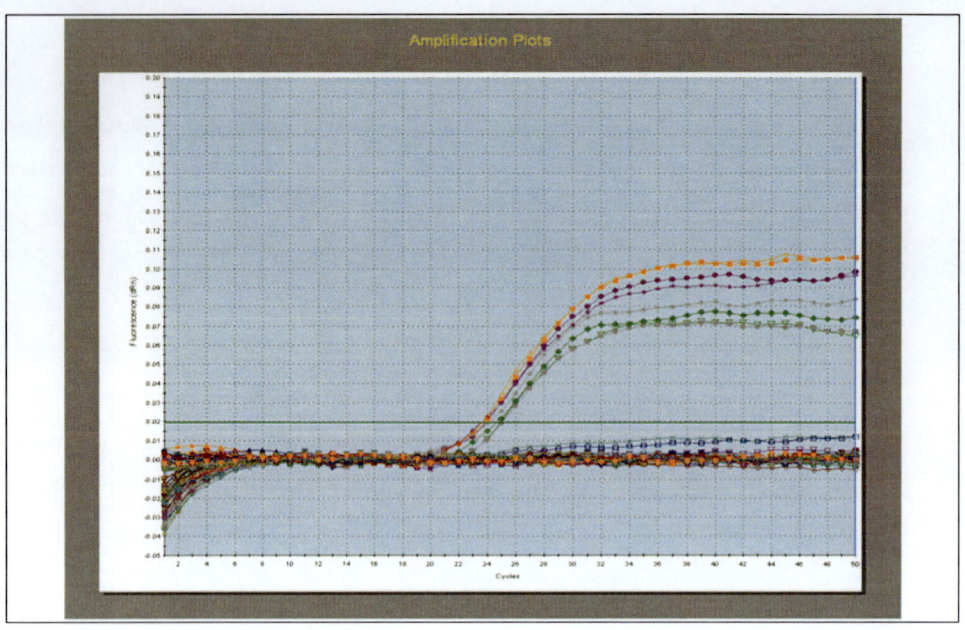

Figure 7: Fluorescence Signal of MLPH (marina blue labeled) in the FAM Filter Channel

The direct comparison of HEX-, Yakima Yellow- and ATTO550-labeled primer-probe-sets showed an advantage of ATTO550 over HEX. It was obvious, that HEX-labeled sets performed erratic over the different concentrations of the standard RNA concentration curve (Figure 8).

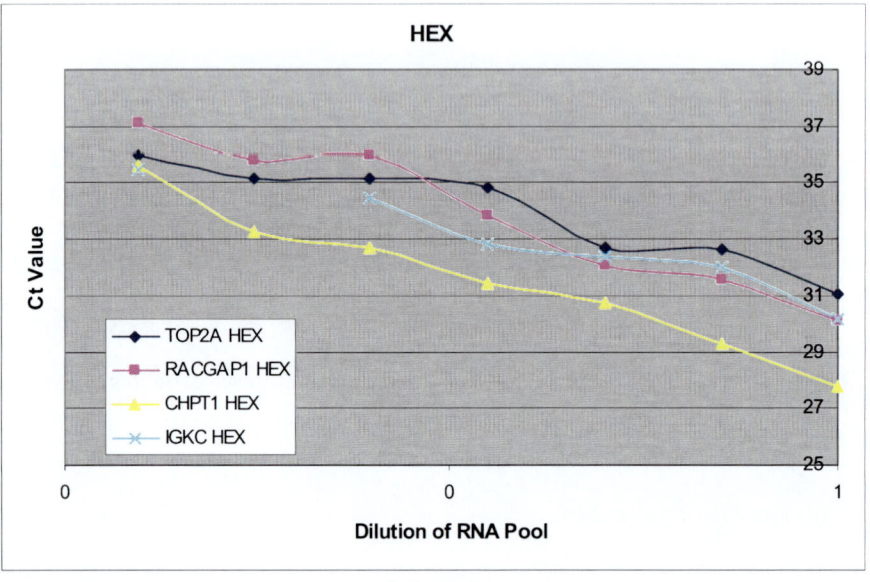

Figure 8: Performance of 4 HEX-Labeled Primer-Probe-Sets on Standard-RNA-Dilution-Series

By taking the mean of the 3-fold determination and plotting the results against a logarithmic scale of RNA-concentrations, the advantages of ATTO550 became obvious.

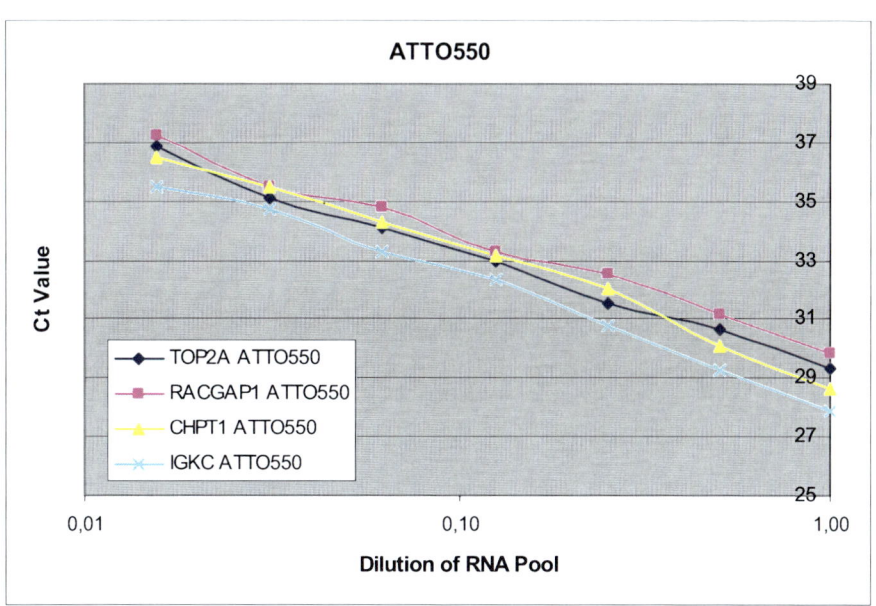

Figure 9: Performance of 4 ATTO550-labeled Primer-Probe-Sets on Standard-RNA-Dilution-Series

Since ATTO550 and Yakima Yellow both performed very well (Figure 9 and 10) but used the same MX3005p-filter set, only one of them could be used for further triplex testing.

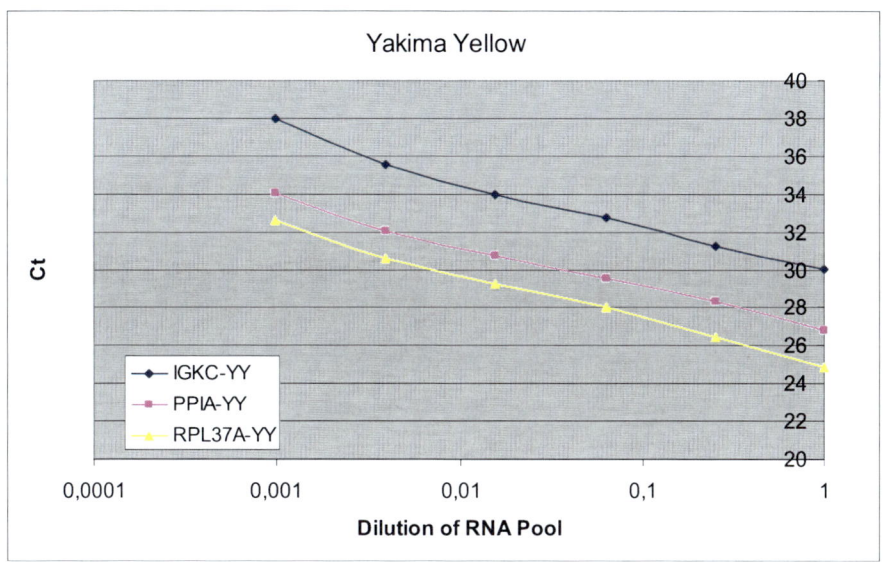

Figure 10: Performance of 3 Yakima Yellow-labeled Primer-Probe-Sets on Standard-RNA-Dilution-Series

The analysis of the tested efficiencies of both primer-probe sets showed, that ATTO550-labeled sets possess a significant lower efficiency than Yakima Yellow-dyes (Table 4). Further testing of ATTO550 was resigned at this point.

Table 4: Comparison of Efficiencies between Singleplex Assays of Yakima Yellow and ATTO550 Labeled Primer-Probe Sets

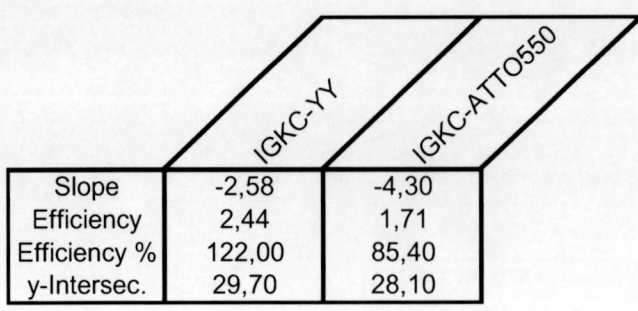

	IGKC-YY	IGKC-ATTO550
Slope	-2,58	-4,30
Efficiency	2,44	1,71
Efficiency %	122,00	85,40
y-Intersec.	29,70	28,10

The remaining dyes – FAM, Cy5, Yakima Yellow and ROX - showed good results and were chosen for further testing. Table 5 presents an overview of the results of the dye comparisons and lists the dyes recommended by Stratagene and the dyes that were additionally tested.

Table 5: Overview of Dye Testing Results and Manufacturers Recommendations

Dye Name	Recommended by Stratagene	Suitable for this Study
FAM	yes	yes
Yakima Yellow	no	yes
Cy5	yes	yes
Marina Blue	no	no
Alexa 405	yes	no
Pacific Blue	no	no
ATTO620	no	no
HEX	yes	yes
Alexa 350	no	no
ROX	yes	yes
ATTO550	no	yes

3.2 Evaluation of Primer-Probe-Performance

One purpose of evaluating the primer-probe-performance is the testing of newly designed or re-ordered primer-probe-sets in order to evaluate their usability during assays and their comparability with former charges. Here, the set was tested on a standard dilution series of RNA and efficiency should range from 90% to 110% to ensure a stable duplication of cDNA during PCR.

Since numerous experiments were performed, only the results of primer-probe sets and dye combinations are mentioned, that showed good efficiencies within the given range and therefore passed all testing. Table 6 gives an overview over the efficiencies of primer-probe sets in singleplex assays used in this study.

Table 6: Overview of Efficiencies of Primer-Probe Sets

Primer-Probe Set	Efficiency (%)
CXCL13-YY	100,00
CHPT1-ROX	91,95
PGR-FAM	99,72
ESR1-YY	88,60
CALM2-Cy5	91,82
MMP1-FAM	94,53
RACGAP1-Cy5	92,04
GAPDH-FAM	87,87
TOP2A-YY	87,99
OAZ1-FAM	90,25
IGKC-YY	90,78
CHPT1-Cy5	91,59
MLPH-FAM	95,65
GAPDH-FAM	87,87

3.3 Identification of Suitable Duplex Combinations

The further testing of primer-probe performance was during comparative assays between singleplex- and duplex-assays. To evaluate whether a primer-probe set is qualified for multiplex assays, several duplex assays with two different genes – each labeled with a different reporter dye – were tested against each other and performances and curves were compared to those with the corresponding individual dye. The duplex assay and the two congruent singleplex assays were always performed on the same 96-well PCR-plate under identical conditions. Table 7 presents an overview of the numerous singleplex-duplex-comparisons made during this study.

Table 7: Comparative Overview of 16 Different Primer-Probe Sets Tested Against Each Other in Duplex kPCR-Reactions (green = Efficiency Difference < 10%, red = Efficiency Difference > 10%, yellow = Uncertain Results)

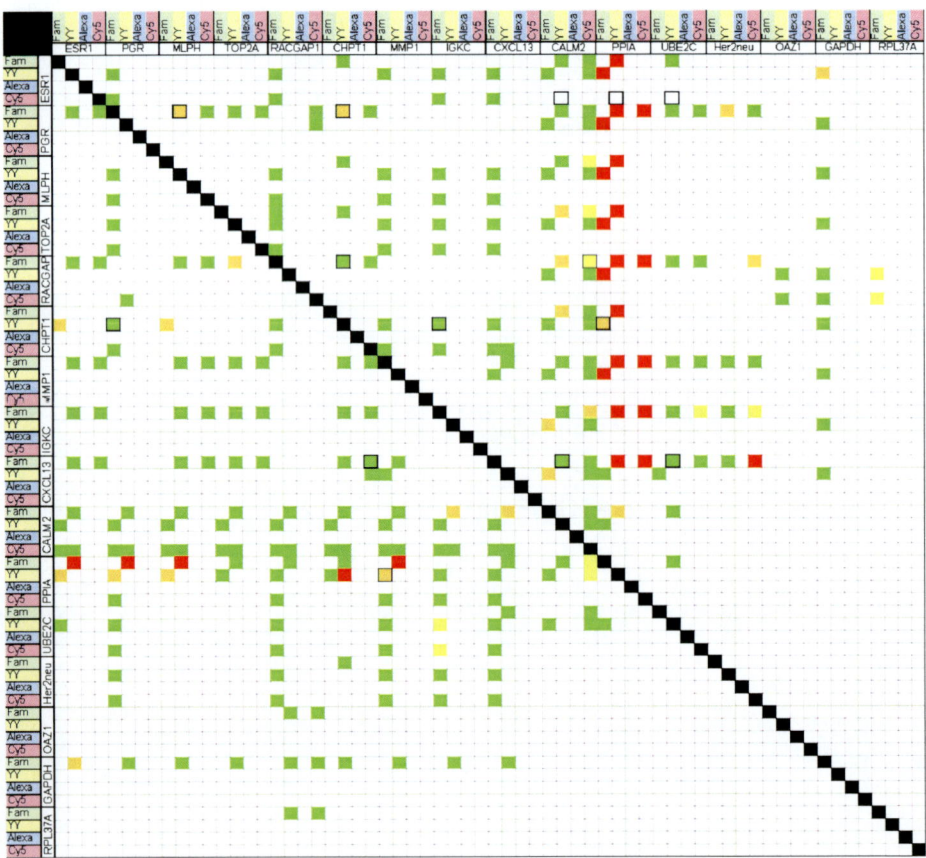

An exemplary combination of two primer-probe-sets that performed well in both single- and duplex assay is CALM2-Cy5 and ESR1-YY (Table 5). It was shown, that the presence of a second oligonucleotide-set did not influence the performance of either one. To determine the individual Ct-values, first the assay was analyzed using the HEX-filter (which is the appropriate choice for the Yakima Yellow dye) and then the Cy5-filter. To avoid crosstalk results between the different filter-channel, it is very important, that wells must not produce Ct-values, when being analyzed with filters for dyes they do not contain.

After this initial crosstalk-check, the performance of the individual single- and duplex-assays was plotted against each other.

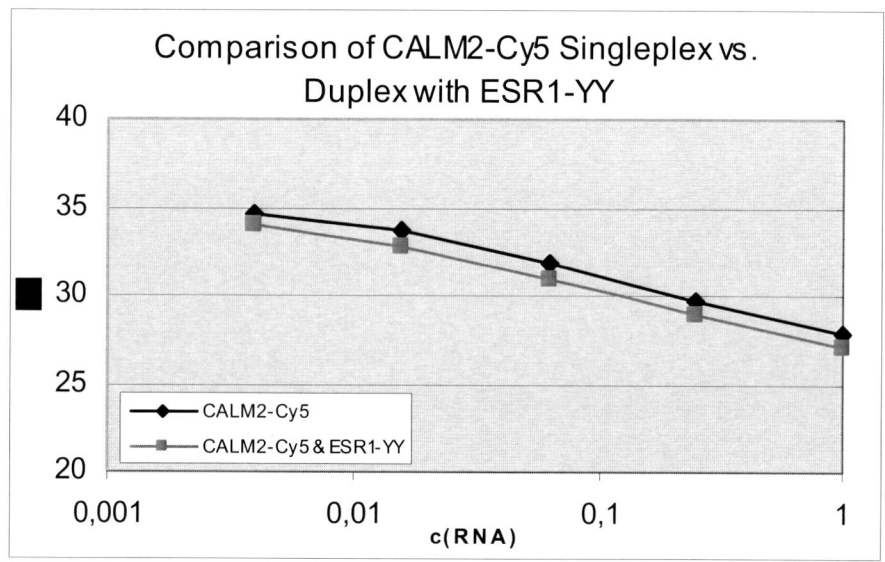

Figure 11: Comparison of Cy5-labeled Primer-Probe-Set (CALM2) in Singleplex Assay vs. Duplex-Assay with Yakima Yellow-labeled Set (ESR1), c(RNA) = Relative Dilution of RNA

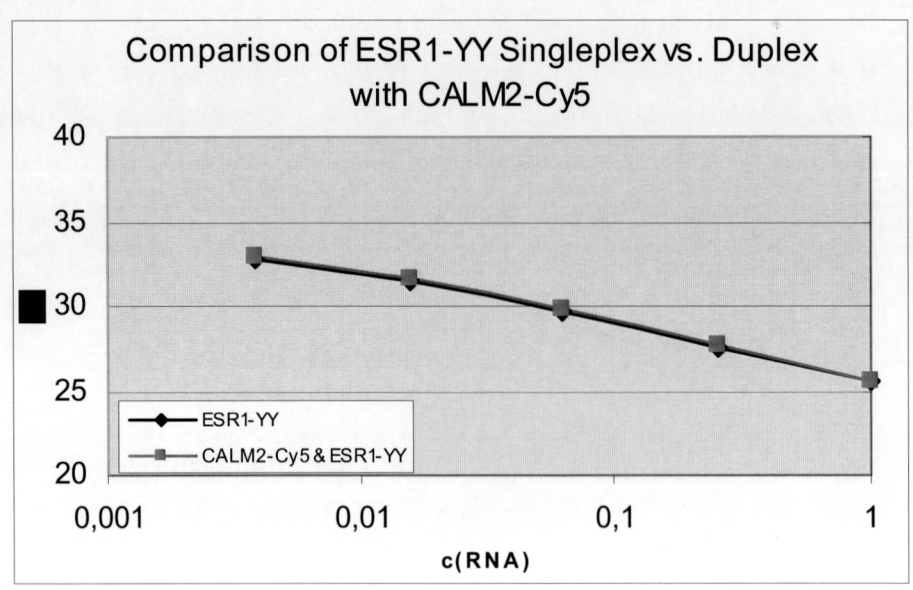

Figure 12: Comparison of Yakima Yellow-Labeled Primer-Probe-Set (ESR1) in Singleplex Assay vs. Duplex-Assay with Cy5-Labeled Set (CALM2) , c(RNA) = Relative Dilution of RNA

CALM2-Cy5 showed slightly lower Ct-Values in the duplex-assay, but since this difference occured likewise at all dilutions (Figure 11) and averages only 0.6 Ct, this will not influence the performance of this primer-probe-duplex compared to the singleplex. ESR1-Yakima Yellow even outperformed CALM2 since singleplex- and duplex-assay did not show any differences at all (Figure 12).

Furthermore the direct comparison of the statistical data focused on the percental efficiency between both singleplexes to the corresponding duplexes showed no relevant differences and led to a green marking in the comparative overview table (Table 7 and 8).

Table 8: Comparison of Efficiencies Between Singleplex Assay of CALM2-Cy5 and ESR1-Yakima Yellow and the Corresponding Duplex

	CALM2-Cy5	ESR1-YY & CALM2-Cy5	ESR1-YY	ESR1-YY & CALM2-Cy5
Slope	-3,29	-3,35	-3,46	-3,29
Efficiency	2,01	1,99	1,95	2,01
Efficiency %	101,33	98,87	94,58	101,48
y-Intersec.	24,72	24,88	26,70	26,57

The results from the duplex testing provided primary information about the compatibility of different primer-probe-sets and individual dyes. At this point, no essential decisions were made concerning possible relinquishments of sets, but to help finding promising combinations for subsequent triplex testing.

3.4 *Identification of Suitable Triplex Combinations*

Testing of triplex combinations of three different genes and dyes was performed similar to the duplex tests. Again, different primer-probe-sets, each labeled with a discrete dye, were tested and the individual performance was compared to the performance of the corresponding singleplex.

An exemplary analysis and a demonstrative illustration of how different primer-probe-sets can perform becomes apparent, when comparing two similar triplex assays in parallel. The triplex TOP2A-YY & MLPH-Cy5 & RACGAP1-FAM and the similar triplex TOP2A-YY & MLPH-FAM & RACGAP1-Cy5 only varied in the dye-assignment of two primer-probe-sets. The first combination with RACGAP1 being FAM-labeled performed very well (Figure 13). The difference between the same dyes compared between single- and triplex amounts to less than 0.5 Ct and showed no change of slope.

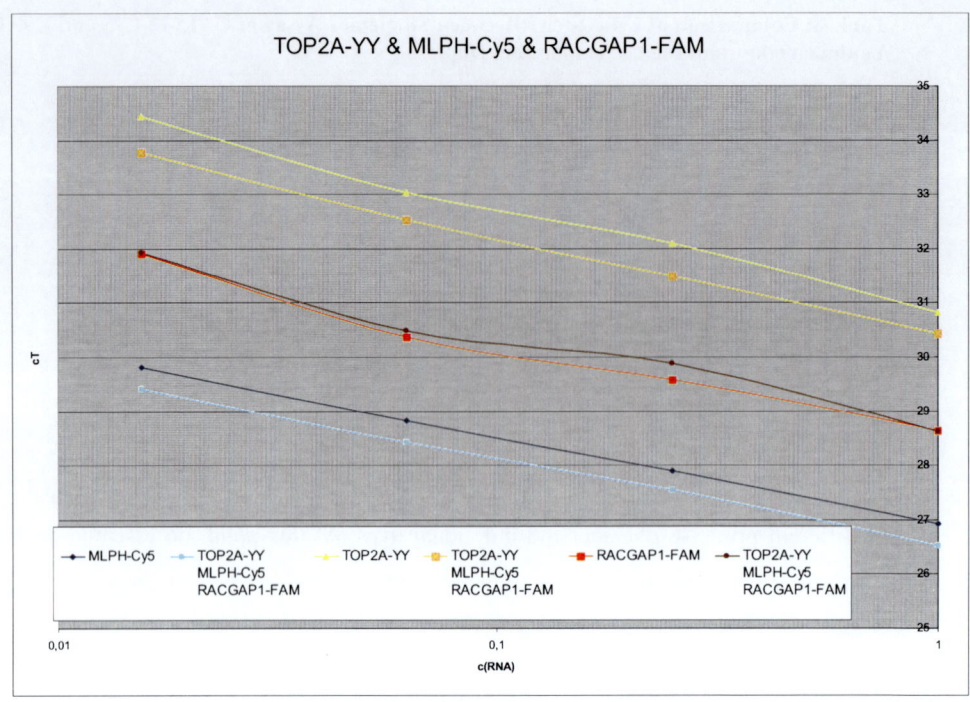

Figure 13: Comparison of Performances of Singleplex Assays with Three Different Reporter Dyes [Yakima Yellow, Cy5, FAM] with the Individual Performances of Dyes within Duplex Assays, c(RNA) = Relative Dilution of RNA

The triplex TOP2A-YY & MLPH-FAM & RACGAP1-Cy5 quantified the identical genes and used the same dyes, whereas the assignment of FAM and Cy5 was interchanged. It was confirmed, that TOP2A-Yakima Yellow performed very similar to the previous triplex, since this primer-probe-set is identical (Figure 14). RACGAP1-Cy5 performed slightly worse, but the difference between the corresponding single- and triplexes was never higher than 1 Ct. MLPH-FAM attracts negative attention, because here the difference between single- and triplex-Ct partially differed more than 2 Ct.

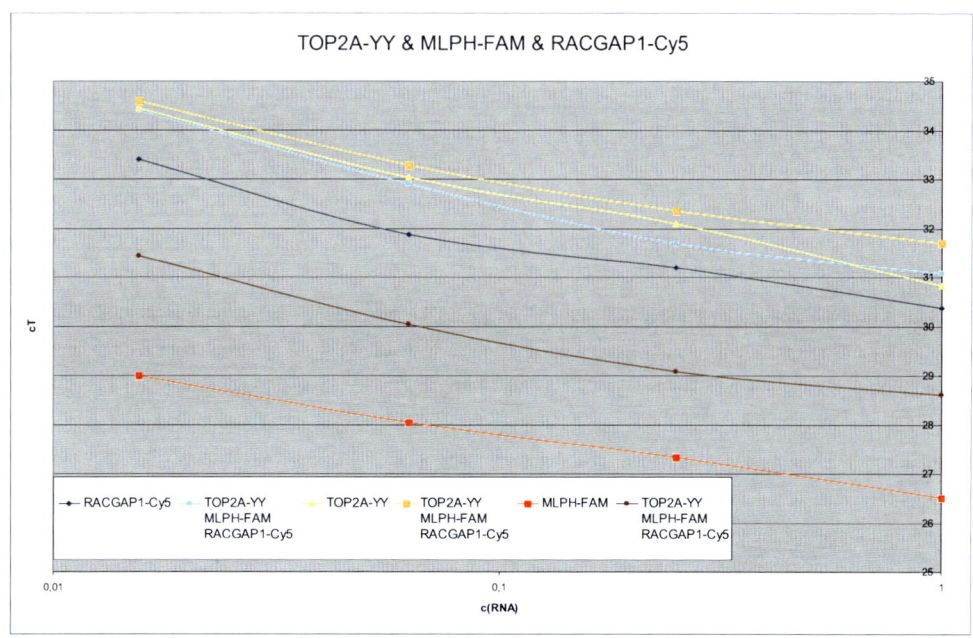

Figure 14: Comparison of Performances of Singleplex Assays with Three Different Reporter Dyes [Yakima Yellow, Cy5, FAM] with the Individual Performances of Dyes within Duplex Assays, c(RNA) = Relative Dilution of RNA

Since both triplexes cover the same three genes of interest, decisions were made in favor of the first triplex due to its better similarity to the corresponding singleplexes.

Besides the difference of Ct-values, great attention was put on the efficiency of the primer-probe-sets and their corresponding performance within a triplex. The more contiguous two compared values were the less influence was exerted on a primer during triplex assays. The following results display the best oligonucleotide-dye combinations that emerged from the numerous testings. Again, great attention was put on the fact, that wells must not produce Ct-values, when being analyzed with filters for dyes they do not contain. The tables are titled with the name of the relevant triplex and show the individual performances of primers within a specific filter-setting.

Table 9: Overview of Triplex Testings - Comparing the Efficiencies (%) of Singleplexes vs. Triplexes

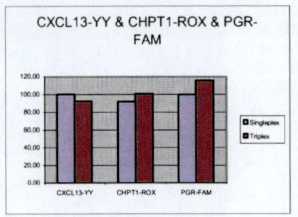

CXCL13-YY & CHPT1-ROX & PGR-FAM	Singleplex	Triplex
CXCL13-YY	100,00	92,51
CHPT1-ROX	91,95	100,70
PGR-FAM	99,72	115,74

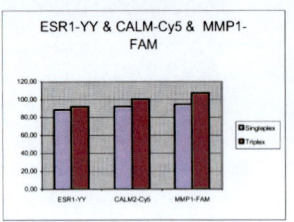

ESR1-YY & CALM-Cy5 & MMP1-FAM	Singleplex	Triplex
ESR1-YY	88,60	92,01
CALM2-Cy5	91,82	100,32
MMP1-FAM	94,53	107,00

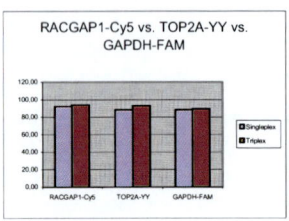

RACGAP1-Cy5 vs. TOP2A-YY vs. GAPDH-FAM	Singleplex	Triplex
RACGAP1-Cy5	92,04	94,00
TOP2A-YY	87,99	92,65
GAPDH-FAM	87,87	89,13

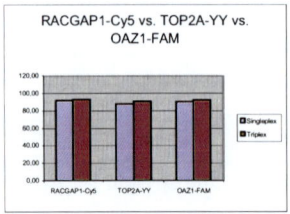

RACGAP1-Cy5 vs. TOP2A-YY vs. OAZ1-FAM	Singleplex	Triplex
RACGAP1-Cy5	92,04	92,88
TOP2A-YY	87,99	90,75
OAZ1-FAM	90,25	92,27

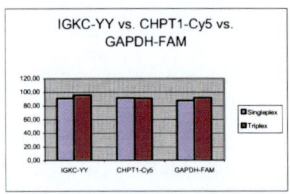

IGKC-YY vs. CHPT1-Cy5 vs. GAPDH-FAM	Singleplex	Triplex
IGKC-YY	90,78	95,54
CHPT1-Cy5	91,59	90,91
GAPDH-FAM	87,87	91,65

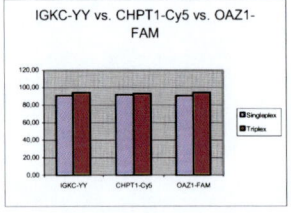

IGKC-YY vs. CHPT1-Cy5 vs. OAZ1-FAM	Singleplex	Triplex
IGKC-YY	90,78	94,03
CHPT1-Cy5	91,59	93,03
OAZ1-FAM	90,25	93,81

In summary, the previous results show, that the initial 12 singleplex assays can be combined into 5 multiplex assays, where some GOIs are quantified in more than 1 assay and with more than one reporter dye:

- CXCL13-YY & CHPT1-ROX & PGR-FAM
- ESR1-YY & CALM2-Cy5 & MMP1-FAM
- IGKC-YY & CHPT1-Cy5 & GAPDH-FAM
- RACGAP1-Cy5 & TOP2A-YY & OAZ1-FAM
- TOP2A-YY & MLPH-FAM & RACGAP1-Cy5

4 Discussion

Within this work it was shown, that singleplex RT-kPCR assays can be combined into multiplex assays. To be able to compare results over a large number of kinetic PCRs and different samples, it was inevitable to keep certain parameters constant (i.e. buffer conditions, reaction volume, usage of identical lots, cycle conditions, etc.). Furthermore, it had to be ensured, that no crosstalk between individual channels occurred.

4.1 Statistical Evaluation of Primer Performances

Understanding the relationship between statistical parameter and the underlying measured data is fundamental for substantiated interpretations. The most weighted parameter observed during this study were the s*lope*, the e*fficiency* and the *y-intersection* of primer-probe sets (Figure 16).

4.1.1 Slope

Theoretically, during PCR every cycle effects a duplication of the template DNA. Therefore, an ideal slope of plotted amounts of nucleic acids vs. amounts of cycles should be -3.32. To comprehend this value, it may by useful to exemplarily calculate a slope with idealistic values. In this instance, we use four dilutions (1:2 respectively) and assume an increase of the Ct-value by 1 per cycle, which correlates to a perfect duplication of DNA [Pfaffl, 2001].

Table 10: Idealistic Ct values and logarithmic display of corresponding dilutions

rel. dilution	abs. dilution	cT	log$_{10}$(dilution)
1	1	20	0,000
1:2	0,5	21	-0,301
1:4	0,25	22	-0,602
1:8	0,125	23	-0,903

By picking any two data points, the ideal slope can now be calculated:

$$slope = \frac{\Delta y}{\Delta x} = \frac{(23-20)}{(-0.903-0)} = \frac{3}{-0.903} = -3.32$$

All calculations are based upon the presumption that the linear model applies.

4.1.2 Efficiency

Like mentioned before, a duplication during every cycle leads to an ideal efficiency of 2. The effective efficiency is calculated by the formula:

$$Efficiency = 10^{(\frac{-1}{slope})}$$

For a better illustration, the efficiency is specified in percentage. For this, it is necessary to subtract 1 from the original efficiency (a perfect duplication per cycle corresponds to an efficiency of 2, which makes 100%). The formula for this is

$$Efficiency(\%) = ((10^{(\frac{-1}{slope})}) - 1) * 100$$

Efficiencies below 100% are particularly reasonable and explainable. There are several factors that lead to inhibitory effects during PCR and especially during multiplex PCR. Since every primer-probe set possesses individual optimal buffer concentrations (esp. $MgSO_4$), temperature optima and cycle conditions, it is obvious that there appear cutbacks for individual sets during a multiplex PCR [Henegariu O, 1997]. Although interactions and hetero-dimerizations should be avoided in the run-up to the experiments, there is no guarantee for this. Erroneous sequences due to SNPs within the sequence of primers may also lead to limitations of primer performances and are tried to be avoided by checking primer-probe-sets for peculiarities within the NCBI SNP database (http://www.ncbi.nlm.nih.gov/sites/entrez?db=snp).

A somewhat rare but existing aspect is an efficiency distinct larger than 100%. Although this should never occur – there are not more than 2 copies emerging from one template during one cycle – there are reactions with 110% efficiency or more. One reasonable explanation for this is a limitation of the reaction within very high concentrated samples. Here, the disposal of free fluorophores is limited by the amount of available reactants resulting in a constant fluorescence signal. Efficiencies clearly above 100% may also be suggestive of methodic errors during the preparation of the dilution series. If one assumes a dilution of 1:2 respectively, but in fact pipettes 1:1.8, the efficiency suffers from that mistake:

$$slope_{err} = \frac{\Delta y_{err}}{\Delta x_{err}} = \frac{(23-20)}{(-1.027-0)} = \frac{3}{-1.027} = -2.921$$

This results in an efficiency of

$$Efficiency_{err}(\%) = ((10^{(\frac{-1}{slope})})-1)*100 = ((10^{(\frac{-1}{-2.921})})-1*100$$

$$= (2.200-1)*100 = 120\%$$

The slope that would be produced by this pipetting error runs clearly below the correct curve.

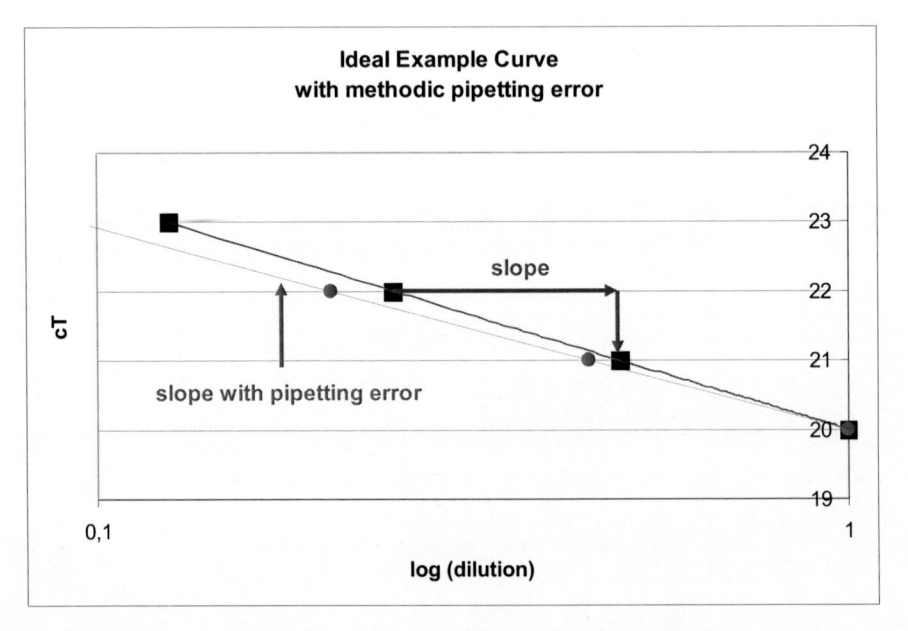

Figure 15: Influence of methodical errors on slope (here: Pipetting error of 10% at every dilution step)

4.1.3 Y-Intercept

The y-intercept of the logarithmic standard curve describes the point, where the regression line meets the y-axis. This point describes the Ct-value of an undiluted sample, and gives an orientation of how good a primer-probe set works when using different dyes on identical samples and concentrations. It was observed, that Ct-values higher then 32 - 34 for the first, undiluted sample will most likely not reach a linear range in diluted samples anymore.

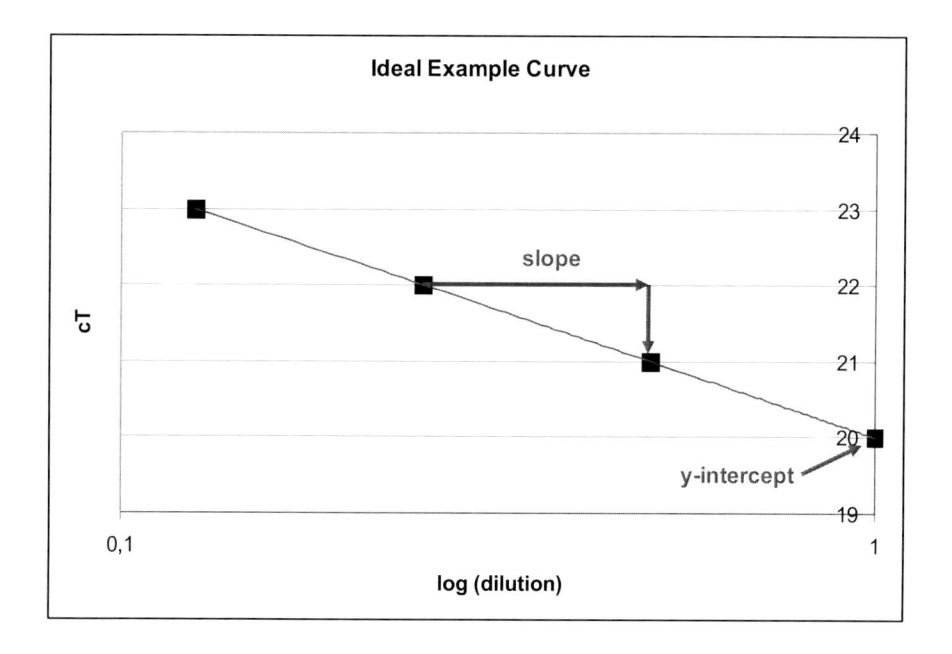

Figure 16: Graphical Illustration of Statistic Parameters Within a Logarithmic Idealized Curve

When comparing the calculated y-intercept of the standard curve with the effective value that was measured, one can get information about the quality of Ct-values from undiluted samples. Since these samples often contain loads of nucleic acid, there might occur some inhibitory effects leading to variations from the expected values. On the other hand, a perfect match between calculated y-intercept and measured data is not a guarantee for a perfect regression curve.

4.2 Suitable Reporter Dyes and Quencher

Finding suitable reporter dyes for multiplex analysis depends on individual requirements for the assay that has to be developed. The present work focused on multiplex analysis of up to three different genes and assays were performed on Stratagenes MX3005p kPCR-system. The by far most important factor when performing multiplex analysis within one tube is the avoidance of crosstalk between the respective wavelengths. After sorting out the dyes, that did not perform with the machine at all, comparative experiments were made to distinguish between different qualities of reporter dyes. Extensive testing was performed to ensure that no single filter set is able to detect fluorescence of another wavelength than the assigned. In fact, the MX3005p kPCR machine showed persuasive results and no crosstalk was detected at all.

One characteristic of the used probes, the quenching with black hole quencher (BHQ) instead of another fluorophore (i.e. TAMRA), may have made a contribution to the fact, that no fluorescence other than the expected was detected. BHQ are able to quench the most common reporter dyes very effectively, because of their ability to quench fluorescence over the entire visible spectrum and into infrared [Bustin, 2002]. A direct comparison between FAM-TAMRA labeled primer probe sets and FAM-BHQ1 labeled sets with identical sequences confirmed that predication: Initial background fluorescence at the beginning and during the run of a kPCR is noticeable lower when BHQ are used..

After all there emerged various points that obviously prevented crosstalk:

- Reliable filter sets with distinct emission- and extinction-wavelengths
- Low background fluorescence due to HPLC purified oligonucleotides
- Optimal quenching of reporter dyes over a large wavelength-range
- No native fluorescence of quencher
- Comprehensive software processing of raw data to subtract background noise

Optimizing the issues mentioned above and going even further by testing more parameter (buffer concentrations, $MgSO_4$ concentrations, temperature profile, etc.) will most probably result in the best multiplex assay possible. This is necessary to compensate some other inevitable weak points. The fact, that this assay is primarily developed for diagnostic purposes requires an easy to handle workflow. Therefore we decided to use single step RT-PCR instead of the more inconvenient but more reliable [Battaglia et al.

1998] two step RT-PCR. One approach to compensate for this taint was the optimization of concentrations of enzymes. We were able to show, that 1.5-fold concentration of both, the reverse transcriptase and the polymerase showed slightly better slopes and lower Ct-values of about 1.

4.3 Optimization of Assays and kinetic RT-PCR Parameter

During this work various adjustment screws for kPCRs emerged during experiments. In the ideal case, the amplification efficiency is 2, that means that DNA amounts double with every cycle run. Experiences show, that this efficiency is rarely achieved, especially when, like in this case, the sample material is limited and other inhibitory effects (like paraffin residues, degraded RNA, etc.) may counteract [Rameckers, 1997]. Most of the GOIs used were expressed high enough to reach Ct values within reasonable ranges. A noticeable fact at the beginning of this study was that the kinetic real time amplification hardly ever reached the saturation. Against the recommendation of Stratagene, who propose 40 cycles during PCR, it was shown, that an increase of cycles up to 50 will deliver better curves. Although this modification does not change the final Ct value, it may help produce more dependable curves and sort out instable runs.

Very intense attention was put on the cross-checking between individual primer probe sets. When using two or more sets, one can observe, that some sets constantly hardly suffer any influence at all while other sets seem to be very accessible for external influences. This requires an intense analysis of multiplex assays and a comprehensive look at the individual sets. While using set A and set B within a duplex reaction, set A may not be influenced at all by set B. But when analyzing set B within its specific filter range, one may often find that inhibitions have taken place. To detect these one-way influences, great attention was put on the comprehensive analysis of data and the single observation of dyes. The inhibitions turned out not to follow any obvious rule. To develop this assay, many combinations had to be tested and individual influences hat to be analyzed. No universal combination between reporter dyes was detected – there did not shape up any rules like "dye A always works best with dye B" or "never use dye B when using dye C".

To approach these influences and to optimize multiplex assays, several actions can be taken to stabilize multiplex performances. A two-fold increase of the enzyme concentration (both polymerase and reverse transcriptase) showed higher Ct-values (0.5 to 1.5 Ct) and slightly more stable curves and performances. This observation could be made for 2 out of 6 tested primer probe sets and was assumed for all further testing. The standard concentration of 0.2 µl was therefore increased up to 0.4µl per reaction (Table 3).

Furthermore it might be necessary to optimize all other parameter influencing a PCR. Like mentioned above, the $MgSO_4$ might influence the performance as well as the specific concentrations of the individual primer and probes. Even the vernier adjustment of the pH-value and cycle temperatures may be expedient, but were not tested during this study.

4.4 Conclusion and Outlook

It was shown, that the number of RT-kPCR reactions could be reduced, by developing multiplex assays and performing parallel analysis of up to three genes within one well. This reduction impacts several important aspects:

- Reduction of cost of approximately 50% for mastermixes per patient
- Higher throughput due to larger sample number per plate
- Larger number and higher variety of GOIs per patient and plate
- Retrenchment of very valuable RNA material
- Generation of resources to run additional quality assurance (housekeeping-genes)

Optimizations emerged to be a product of several factors: PCR parameter, buffer, choice of reporter dyes and their combinations, $MgSO_4$ addition, and many more. Although experiments based upon an existing platform, substantial testing is inevitable due to the unpredictable interaction of primer probe sets within multiplex assays. The most promising and secure way is a time consuming testing of every possible combination. This approach is indicated, when the emerging assay will be used as a commercial test in the future and bulk amounts will be needed.

The other aspect of extensive testing is a qualification for commercial use. Besides very restrictive company-internal requirements, regulatory authorities will demand evidence to proof the applicability of this test.

However, this work approached this subject from the biomolecular aspect. There is still enough evaluation and validation to be done, until a possible test can be offered as a multiplex assay. The next step to do is to statistically confirm that quantifications of large numbers of patient samples made with this multiplex layout are comparable to the results from corresponding singleplex assays.

5 Summary

The diagnostic panel, that is under development at Siemens Healthcare Diagnostics at Cologne, Germany, targets patients with node-negative breast cancer and predicts the probability of distant metastasis after surgery. Besides ample regulatory issues, cost factors and practical aspects are the main criteria for a successful assay. This work outlines the successful reduction of 12 singleplex kPCR assays to 5 multiplex assays and the corresponding activity on this way.

During this work, several genes with different labeled reporter dyes had to be tested against each other. To perform such comparative testing, it was a basic requirement to possess a standard reference RNA pool that contains different nucleic acid sources and therefore increase the number of possible genes that can be amplified. The standard reference RNA pool used here contained different RNA samples from breast cancer patients of different origin and was combined with RNA from the MCF7 human breast cancer cell line.

First, suitable reporter dyes had to be discovered, that were detectable with the given kPCR system, a Stratagene MX3005p. For this, a small number of primer-probe sets was designed and ordered with eleven different dyes in each case. Several dyes had to be discarded because of crosstalk between other filter sets or because of poor signal intensity. After identifying the best reporter dyes, the complete diagnostic gene panel was ordered with the dyes for further testing.

The following testing was performed by concentrating the main focus of attention upon the performance of a singleplex assay compared to the performance of the same assay within duplex analysis. The performance of primer-probe sets was regarded as both the change of efficiency and change of y-intercepts on plotted result analysis. Specific statistical analysis between corresponding data made it possible to identify optimal combinations.

The identification of good duplexes was followed by an identification of suitable triplexes. The methods for comparison were identical to the previous duplex testing and resulted in 5 triplex-assays that are able to substitute the initial singleplex assays. The next step to do is to statistically confirm that quantifications of large numbers of patient samples made with this multiplex layout are comparable to the results from corresponding singleplex assays.

6 Literature

Battaglia M, Pedrazzoli P, Palermo B, Lanza A, Bertolini F, Gibelli N. Da Prada GA, Zambelli A, Perotti C, Robustelli della Cuna G (1998): *Epithelial tumour cell detection and the unsolved problems of nested RT-PCR*, Nature, 22:693-698

Bustin, SA (2002): *Quantification of mRNA using real-time reverse transcription PCR (RT-PCR): trends and problems*, Journal of Molecular Endocrinology, 2:169-193

Buyse M, Loi S, van't Veer L, Viale G, Delorenzi M, Glas AM, d'Assignies MS, Bergh J, Lidereau R, Ellis P, Harris A, Bogaerts J, Therasse P, Floore A, Amakrane M, Piette F, Rutgers E, Sotiriou C, Cardoso F, Piccart MJ; TRANSBIG Consortium (2006): *Validation and clinical utility of a 70-gene prognostic signature for women with node-negative breast cancer*, Journal of the National Cancer Institute 98:1183-1192

Eurogentec: *Custom Oligonucleotides*, Belgium, 2007

Freeman WM, Walker SJ, Vrana KE (1999): *Quantitative RT-PCR: Pitfalls and Potential*, BioTechniques, 26:112-125

Ganz PA, Kwan L, Stanton AL, Krupnick JL, Rowland JH, Meyerowitz BE, Bower JE, Belin TR (2004): *Quality of life at the end of primary treatment of breast cancer: first results from the moving beyond cancer randomized trial*, Journal of the National Cancer Institute, Journal of the National Cancer Institute 96(5):376-387

Garcia-Closas M, Hall P, Nevanlinna H, Pooley K, Morrison J (2008): *Heterogeneity of Breast Cancer Associations with Five Susceptibility Loci by Clinical and Pathological Characteristics,* Public Library of Science; 4(4): e1000054

Gilboa E, Mitra SW, Goff S, Baltimore D (1979): *A detailed model of reverse transcription and tests of crucial aspects*, Cell, 18(1):93-100

Heid CA, Stevens J, Livak KJ, Williams PM (1996): *Real time quantitative PCR*, Genome Research, 6:986-994

Henegariu O, Heerema NA, Dlouhy SR, Vance GH, Vogt PH (1997): *Multiplex PCR: Critical Parameters and Step-by-Step Protocol*, BioTechniques, 23:504-511

Hennig G, Hildenbrand, K.: *Magnetic Particles with a Closed Ultrathin Silica Layer, Method for the Production thereof and their use*, Patent WO/2006/136314, World Intellectual Property Organization

Marras S (2006): *Selection of Fluorophore and Quencher Pairs for Fluorescent Nucleic Acid Hybridization Probes*, Methods in Molecular Biology - Fluorescent Energy Transfer Nucleic Acid Probes, 1:3-16

Mojica WD, Stein L, Hawthorn L (2008): *Universal Reference RNA is Not a Representative Normal Sample for Oligonucleotide Microarray Studies*, Pathology & Oncology Research, 14(3): 243-251

Mülhardt C (2008): *Der Experimentator: Molekularbiologie*, Spektrum Akademischer Verlag., Heidelberg, 2nd Issue

Mullis KB, Faloona FA (1987): *Specific synthesis of DNA in vitro via a polymerase-catalyzed chain reaction*, Methods in enzymology, 155:335-50

Parkin DM, Bray F, Farlay J, Pisani P (2005): *Global cancer statistics,* Cancer Journal for Clinicians, 55: 74-108

Pfaffl, M (2001): A new mathematical model for relative quantification in real-time RT-PCR, Nucleic Acid Research, 29(9):e45

Rameckers J, Hummel S, Herrmann B (1997): *How many cycles does a PCR need? Determinations of cycle numbers depending on the number of targets and the reaction efficiency factor*, Naturwissenschaften, Springer Berlin / Heidelberg, 84(6): 259-262

American Association for Cancer Research (2007): *Gene Profiling Predicts Resistance To Breast Cancer Drug Herceptin*, American Association for Cancer Research

Stratagene (2005): *Strategies Newsletter*, 18(2)

Stropp U[1], von Toerne C[1], Weber K[1], Schmidt M[2], Gehrmann M[1] (2008): *Measurement of Breast Cancer Prognosis Genes in Paraffin-Embedded, Formalin-Fixed Tumor Tissue*, [1]Siemens Healthcare Diagnostics Inc., Cologne, Germany, [2]Johannes Gutenberg University, Medical School, Mainz, Germany

van 't Veer LJ, Dai H, van de Vijver MJ, He YD, Hart AA, Mao M, Peterse HL, van der Kooy K, Marton MJ, Witteveen AT, Schreiber GJ, Kerkhoven RM, Roberts C, Linsley PS, Bernards R, Friend SH (2002): *Gene expression profiling predicts clinical outcome of breast cancer*, Nature, 415(6871):530-6

Wink M (2004): *Molekulare Biotechnologie – Konzepte und Methoden*, WILEY-VCH Verlag GmbH

Wolff A, Hammond E, Schwartz JN, Hagerty KL, Allred DC, Cote RJ, Dowsett M, Fitzgibbons PL, Hanna WM, Langer A, McShane LM, Paik S, Pegram MD, Perez EA, Press MF, Rhodes A, Sturgeon C, Taube SE, Tubbs R, Vance GH, van de Vijver M, Wheeler TM, Hayes DF (2007): *American Society of Clinical Oncology/College of American Pathologists Guideline Recommendations for Human Epidermal Growth Factor Receptor 2 Testing in Breast Cancer*, Journal of Clinical Oncology, 25(1):118-145

7 Appendix

7.1 *Abbreviations*

BHQ	Black Hole Quencher
cDNA	Complementary Deoxyribonucleic Acid
Ct	Cycle Threshold
DNA	Deoxyribonucleic Acid
ER	Estrogen Receptor
FFPE	Formalin-Fixed Paraffin-Embedded
FRET	Fluorescence Resonance Energy Transfer
GOI	Gene of Interest
kPCR	kinetic PCR
µl	Microliter
µM	Micromol
nm	Nanometer
NA	Nucleic Acid
nM	Nanomol
PCR	Polymerase Chain Reaction
QPCR	Quantitative Polymerase Chain Reaction
RNA	Ribonucleic Acid
RT	Room Temperature
RT-PCR	Reverse-Transcriptase Polymerase Chain Reaction
YY	Yakima Yellow (dye)

7.2 Figures

Figure 1: Schematic Workflow of Nucleic Acid Extraction from FFPE Samples 8

Figure 2: Thermal Profile of RT-QPCR on the Stratagene Mx3005p 12

Figure 3: Absorption Spectra of FAM (green), Yakima Yellow (orange), ROX (red) and Cy5 (blue), Source: http://www.biosearchtech.com/hot/multiplexing.asp 15

Figure 4: Emission Spectra of FAM (green), Yakima Yellow (orange), ROX (red) and Cy5 (blue), Source: http://www.biosearchtech.com/hot/multiplexing.asp 15

Figure 5: Comparative evaluation of GAPDH-FAM in duplex reactions with RACGAP1-Cy5 (marked magenta) and with RACGAP1-YY (marked yellow) versus the corresponding singleplex reaction of GAPDH-FAM (marked blue). 16

Figure 6: Fluorescence signal of MLPH (marina blue labeled) in the Alexa 350 filter channel 19

Figure 7: Fluorescence signal of MLPH (marina blue labeled) in the FAM filter channel 20

Figure 8: Performance of 4 HEX-labeled Primer-Probe-Sets on Standard-RNA-Dilution-Series 20

Figure 9: Performance of 4 ATTO550-labeled Primer-Probe-Sets on Standard-RNA-Dilution-Series 21

Figure 10: Performance of 3 Yakima Yellow-labeled Primer-Probe-Sets on Standard-RNA-Dilution-Series 21

Figure 11: Comparison of Cy5-labeled Primer-Probe-Set (CALM2) in Singleplex Assay vs. Duplex-Assay with Yakima Yellow-labeled Set (ESR1) 25

Figure 12: Comparison of Yakima Yellow-labeled Primer-Probe-Set (ESR1) in Singleplex Assay vs. Duplex-Assay with Cy5-labeled Set (CALM2) 26

Figure 13: Comparison of Performances of Singleplex Assays with Three Different Reporter Dyes [Yakima Yellow, Cy5, FAM] with the Individual Performances of Dyes within Duplex Assays 28

Figure 14: Comparison of Performances of Singleplex Assays with Three Different Reporter Dyes [Yakima Yellow, Cy5, FAM] with the Individual Performances of Dyes within Duplex Assays 29

Figure 15: Influence of methodical errors on slope (here: Pipetting error of 10% at every dilution step) 34

Figure 16: Graphical illustration of statistic parameters within a logarithmic idealized curve 35

7.3 Tables

Table 1: Dilution Series of Human Breast Cancer Samples for Setting up a Standard Reference RNA Pool.. 9
Table 2: Setup of a Standard QPCR Assay 11
Table 3: Overview of Tested Genes and Their Derivation... 18
Table 4: Comparison of Efficiencies Between Singleplex Assays of Yakima Yellow and ATTO550 Labeled Primer-Probe Sets 22
Table 5: Overview of Dye Testing Results and Manufacturers Recommendations..................... 22
Table 6: Overview of Efficiencies of Primer-Probe Sets 23
Table 7: Comparative Overview of 16 Different Primer-Probe Sets Tested Against Each Other in Duplex QPCR-Reactions (green = Efficiency Difference < 10%, red = Efficiency Difference > 10%, yellow = Uncertain Results) 24
Table 8: Comparison of Efficiencies Between Singleplex Assay of CALM2-Cy5 and ESR1-Yakima Yellow and the Corresponding Duplex 27
Table 9: Overview of Triplex Testings - Comparing the Efficiencies (%) of Singleplexes vs. Triplexes... 30
Table 10: Idealistic Ct values and Logarithmic Display of Corresponding Dlutions................... 32

7.4 Oligonucleotide Sequences

Gene	Int. Name	Sequence
ESR1	BC170_ESR1	5'-ATGCCCTTTTGCCGATGCA-3'
ESR1	BC170for	5'-GCCAAATTGTGTTTGATGGATTAA-3'
ESR1	BC170rev	5'-GACAAAACCGAGTCACATCAGTAATAG-3'
MLPH	R49_MLPH	5'-CCAAATGCAGACCCTTCAAGTGAGGC-3'
MLPH	R49for	5'-TCGAGTGGCTGGGAAACTTG-3'
MLPH	R49rev	5'-AGATAGGGCACAGCCATTGC-3'
PGR	BC172_PGR	5'-TTGATAGAAACGCTGTGAGCTCGA-3'
PGR	BC172for	5'-AGCTCATCAAGGCAATTGGTTT-3'
PGR	BC172rev	5'-ACAAGATCATGCAAGTTATCAAGAAGTT-3'
TOP2A	R70_TOP2A	5'-CAGATCAGGACCAAGATGGTTCCCACAT-3'
TOP2A	R70for	5'-CATTGAAGACGCTTCGTTATGG-3'
TOP2A	R70rev	5'-CCAGTTGTGATGGATAAAATTAATCAG-3'
CXCL13	R109_Cxcl13	5'-TGGTCAGCAGCCTCTCTCCAGTCCA-3'
CXCL13	R109for	5'-CGACATCTCTGCTTCTCATGCT-3'
CXCL13	R109rev	5'-AGCTTGTGTAATAGACCTCCAGAACA-3'
MMP1	R-MMP1mav1	5'-AGAGAGTACAACTTACATCGTGTTGCGGCTCA-3'
MMP1	MMP1-FWmav1	5'-AGATGAAAGGTGGACCAACAATTT-3'
MMP1	MMP1-REmav1	5'-CCAAGAGAATGGCCGAGTTC-3'
IGKC	R61_IgKC	5'-AGCAGCCTGCAGCCTGAAGATTTTGC-3'
IGKC	R61for	5'-GATCTGGGACAGAATTCACTCTCA-3'
IGKC	R61rev	5'-GCCGAACGTCCAAGGGTAA-3'
CALM2	R117-CALM2	5'-TCGCGTCTCGGAAACCGGTAGC-3'
CALM2	R117for	5'-GAGCGAGCTGAGTGGTTGTG-3'
CALM2	R117rev	5'-AGTCAGTTGGTCAGCCATGCT-3'
PPIA	R115-PPIA	5'-TGGTTGGATGGCAAGCATGTGGTG-3'
PPIA	R115for	5'-TTTCATCTGCACTGCCAAGACT-3'
PPIA	R115rev	5'-TATTCATGCCTTCTTTCACTTTGC-3'
CHPT1	R138_CHPT1	5'-CCACGGCCACCGAAGAGGCAC-3'
CHPT1	R138for	5'-CGCTCGTGCTCATCTCCTACT-3'
CHPT1	R138rev	5'-CCCAGTGCACATAAAAGGTATGTC-3'
RACGAP1	R125-2_RACGAP1	5'-ACTGAGAATCTCCACCCGGCGCA-3'
RACGAP1	R125-2for	5'-TCGCCAACTGGATAAATTGGA-3'
RACGAP1	R125-2rev	5'-GAATGTGCGGAATCTGTTTGAG-3'

PAEP	Sc88_PAEP	5'-AAGCCCTCAGCCCTGCTCTCCATC-3'
PAEP	Sc88for	5'-CACAGAATGGACGCCATGAC-3'
PAEP	Sc88rev	5'-AAACCAGAGAGGCCACCCTAA-3'
UBE2C	R65_UBE2C	5'-TGAACACACATGCTGCCGAGCTCTG-3'
UBE2C	R65for	5'-CTTCTAGGAGAACCCAACATTGATAGT-3'
UBE2C	R65rev	5'-GTTTCTTGCAGGTACTTCTTAAAAGCT-3'
ERBB2	FPE044_ERBB2	5'-CAGATTGCCAAGGGGATGAGCTACCTG-3'
ERBB2	FPE044for	5'-CCAGGACCTGCTGAACTGGT-3'
ERBB2	FPE044rev	5'-TGTACGAGCCGCACATCC-3'
RPL37A	R16_RPL37A	5'-TGGCTGGCGGTGCCTGGA-3'
RPL37A	R16for	5'-TGTGGTTCCTGCATGAAGACA-3'
RPL37A	R16for	5'-GTGACAGCGGAAGTGGTATTGTAC-3'
OAZ1	BC268-RNA_OAZ1	5'-TGCTTCCACAAGAACCGCGAGGA-3'
OAZ1	BC268for	5'-CGAGCCGACCATGTCTTCAT-3'
OAZ1	BC268rev	5'-AAGCCCAAAAAGCTGAAGGTT-3'
GAPDH	FPE029_GAPDH	5'-AAGGTGAAGGTCGGAGTCAACGGATTTG-3'
GAPDH	FPE029for	5'-GCCAGCCGAGCCACATC-3'
GAPDH	FPE029rev	5'-CCAGGCGCCCAATACG-3'

7.5 Origins of Breast Cancer Samples for MAVPOOL080623a

a) Prof. Dr. med. Störkel, Helios Klinikum Wuppertal
b) Prof. Dr. Carsten Denkert, Universitätsklinikum Berlin
c) MCF7 cell line

Continuous Number	Internal Number	Source	Continuous Number	Internal Number	Source
1	52.26	a)	42	69.03	a)
2	52.28	a)	43	69.04	a)
3	52.38	a)	44	69.06	a)
4	52.41	b)	45	69.07	a)
5	52.53	c)	46	69.08	a)
6	54.01	b)	47	69.09	a)
7	54.02	b)	48	69.10	a)
8	54.04	c)	49	69.12	b)
9	54.07	b)	50	72.02	a)
10	54.08	b)	51	72.03	a)
11	54.14	b)	52	72.04	a)
12	54.15	b)	53	72.06	a)
13	55.01	b)	54	72.08	a)
14	55.02	b)	55	72.09	a)
15	55.03	b)	56	74.08	b)
16	55.06	b)	57	74.09	b)
17	55.07	b)	58	74.14	b)
18	55.08	b)	59	83.02	b)
19	57.01	b)	60	83.04	b)
20	59.04	b)	61	83.06	b)
21	59.09	b)	62	83.07	b)
22	59.22	c)	63	83.11	b)
23	61.01	b)	64	86.10	b)
24	61.03	b)	65	86.11	b)
25	61.05	b)	66	86.12	b)
26	61.07	b)	67	86.13	b)
27	61.09	b)	68	86.14	b)
28	63.07	b)	69	86.15	b)
29	63.10	b)	70	86.16	b)
30	63.16	b)	71	93.02	b)
31	65.16	b)	72	93.03	b)
32	67.01	b)	73	93.04	b)
33	67.04	b)	74	93.05	b)
34	67.08	b)	75	93.07	b)
35	67.11	b)	76	93.08	b)
36	67.12	b)	77	93.13	b)
37	68.01	b)	78	93.14	b)
38	68.02	b)	79	94.02	b)
39	68.06	b)	80	94.03	b)
40	68.07	b)	81	94.09	b)
41	69.02	a)	82	94.10	b)
			83	94.16	b)

7.6 Acknowledgements

I would like to thank Dr. Christoph Petry, Head of Molecular Research Germany at Siemens Healthcare Diagnostic Products GmbH, Cologne, Germany, for the possibility to accomplish these studies and his substantial support at all points during the past years.

Also I would like to thank my boss and head of the laboratory Dr. Udo Stropp for his continuous support and professional help in this time. His courtesy and support made everything a lot easier.

Furthermore I want to thank Priv.-Doz. Dr. Ralf Kronenwett for proofreading this work and for his invaluable advices. With his comprehensive experience and both objective and considerate assistance, I found a great help for writing this work.

Last but not least I want to thank my dear colleague Dipl.Biol. Corinna Kik, who also kept a check on this work and provided helpful advice. In addition to this, one cannot have a more helpful and amicable workmate and our teamwork, especially during these comprehensive testing, was outstanding.